The Zeppelin Story

The Zeppelin Story

John Christopher

The History Press

Also in this series:

Half title page: *A new generation of Zeppelins – the prototype NT07 001 on the mobile mooring mast beside the hangar at Friedrichshafen in southern Germany.*

Title page: *A superb image of a pre-WWI Zeppelin. (US Library of Congress)*

Published in the United Kingdom in 2010 by
The History Press
The Mill · Brimscombe Port · Stroud · Gloucestershire · GL5 2QG

British Library Cataloguing in Publication Data
A catalogue record for this book is available from the British Library.

Hardback ISBN 978-0-7524-5175-6

Typesetting and origination by The History Press
Printed in Italy

NT07 003 during routine maintenance at Friedrichshafen.

CONTENTS

ACKNOWLEDGEMENTS

Most of the images in this book have either come from the archives of Airship Initiatives Ltd or are new photographs taken by the author. Others are from the US Library of Congress, Goodyear, the US Navy, the US National Oceanic and Atmospheric Administration (NOAA), CargoLifter, Hybrid Air Vehicles (ATG), 21st Century Airships and Zeppelin Tourismus Forderverein.

I would also like to thank my wife, Ute, for her continued support and many hours of proofreading, and my children Anna and Jay.

The quotations in this book come from a number of sources including: *My Zeppelins* by Hugo Eckener; *The Log of HMA R34 – Journey to America and Back* by Air Commodore E.M. Maitland; *Transatlantic Airships* by John Christopher; *The Zeppelins* by Captain Ernst Lehmann and Howard Mingos; *Airshipwreck* by Len Deighton and Arnold Schwartzman; *Dr Eckener's Dream Machine* by Douglas Botting; *The War in the Air* by H.G. Wells; *Trail Blazing in the Skies* by Goodyear Tyre & Rubber Company; *The Zeppelin Reader* edited by Robert Hedin; plus *Future Life*, *Telegraph Travel*, *Modern Transport* and *The American Magazine*.

This book is dedicated to the memory of Roger Munk and George Spyrou, two fine airshipmen who died while it was being written. Roger was an inspiration to many as the creator of the first modern airship with the Skyship series in the 1980s, and it was George's company that kept them flying into the twenty-first century.

> If airships are to be of any real use for military purposes, it is imperative that these airships must be able to navigate against very strong air currents; they must be able to remain in the air without landing for at least twenty-four hours, so that they can perform really long reconnoitring tours … In other words: large airships will be needed.
>
> Count Ferdinand von Zeppelin, Report to the King of Württenberg, 1887

The Zeppelins were born out of warfare. Their creator, Count Ferdinand von Zeppelin, had envisioned a fleet of aerial behemoths emblazoned with the imperial eagle of the German emperor to ensure the nation's domination of the skies. Each one would be capable of navigating against the elements on long duration flights of more than twenty-four hours. They would be able to carry heavy loads of men, supplies and ammunition or bombs. In other words, they needed to be big. Very big indeed.

'If it is possible to solve these problems,' Count Zeppelin reported to the King of Württenberg in 1887, 'the importance of airships in the future will certainly be immeasurable. Not only will they become important in warfare; they will be used for civil transportation… They will also be used on expeditions of discovery (to the North Pole, to central Africa).' These prophecies were extraordinary, especially as they came from an elderly Prussian cavalry officer who had left the imperial army under something of a cloud. Even more

This monument to Count Ferdinand von Zeppelin, 1838–1917, bears the motto: 'One only has to have the will for it to succeed.'

➤ *The Montgolfier brothers' balloons caused widespread wonderment when they first flew in France in 1783, but the hot-air balloons soon gave way to the more robust hydrogen variety.*

➤➤ *This early nineteenth-century engraving reveals the wide variety of visionary schemes for balloons and airships ranging from flying doughnuts to fish-shaped craft.*

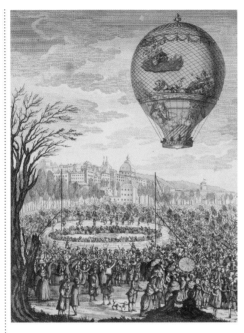

so considering that they were made at a time when only a handful of experimental airships had appeared on the scene. This was seven years before the Wright brothers so famously made their first heavier-than-air hops at Kittyhawk, and until then the sky remained the province of a small band of the upwardly-mobile: the balloonists.

Zeppelin had first encountered tethered observation balloons when he travelled to the USA to act as an official observer during the American Civil War, and he must also have been aware of the free-flying balloons used by the French to escape the Prussian's siege of Paris in 1870–71. Balloons, by their very nature, can only travel with the wind which severely limits their ability to manoeuvre above specific targets. During the nineteenth century they remained, with the exception of a handful of scientifically-minded high-fliers, the plaything of the travelling showmen who entertained the paying public with their dazzling aerial

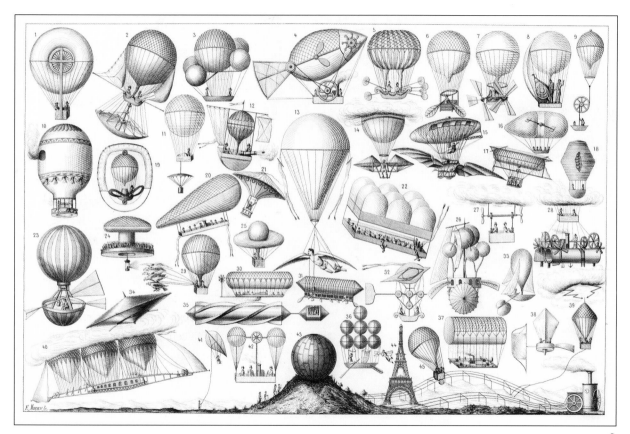

9

displays. Inevitably, the novelty of a balloon launch began to wear thin and, as interest dwindled, the showmen resorted to ever more exotic stunts. They ascended beneath their balloons astride horses, descended on precarious parachutes and, heaven forbid, they even resorted to sending their wives aloft. None of this served the cause of the visionaries who continued in their quest to find the means of propulsion and steering to make them into a viable military tool, to create what the French term a 'dirigible' balloon.

As early as 1784, just a year after the Montgolfier brothers' fragile hot-air balloons had first flown, a French army engineer named Jean Baptiste Marie Meusnier had devised an egg-shaped airship to be inflated with hydrogen gas. This was, in essence, what is now known as a 'pressure' or 'non-rigid' airship as it had no internal framework or structure and its shape was to be maintained by internal pressure alone. Meusnier had even solved the problem of controlling the internal gas pressure despite changes to the external ambient air pressure (see Chapter 10). But what was lacking was an efficient means of propulsion.

In 1852 another Frenchman, Henri Gifford, tackled the propulsion issue with steam power, and although his airship is reported to have attained speeds of up to 6mph (9.7km/h), the wisdom of placing a boiler in such close proximity to the flammable hydrogen was highly questionable. Throughout the remainder of the nineteenth century a veritable army of inventors and engineers kept the patent offices occupied with all sorts of methods of propelling an airship, including assorted human-driven flapping devices and paddle wheels. In 1883 Gaston and Albert Tissandier flew an electrically powered airship, and

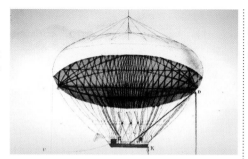

◀ Le Meusnier's prophetic 1784 design for a non-rigid airship. Never built, it did portray the shape of things to come.

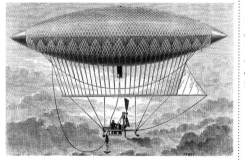

◀ Frenchman Henri Giffard's steam-powered airship of 1852. Note that the exhaust from the boiler points downwards away from the hydrogen-filled envelope.

the following year a pair of engineers in the French army, Charles Renard and Arthur Krebs, developed a lightweight battery for

▲ *The dapper Brazilian aeronaut, Alberto Santos-Dumont, delighted turn-of-the-century Paris with his aerial exploits aboard his one-man runabouts.*

➤ *In 1901 Alberto Santos-Dumont won the Deutsch Prize by completing a circular course across Paris and around the Eiffel Tower.*

their *La France* airship. But in neither case was the airspeed sufficient to overcome even the slightest headwind.

The answer to this predicament came in the form of the internal combustion engine, in particular through the work of Gottlieb Daimler. Its application to airships was brought to public attention through the activities of the wealthy and dapper Brazilian-born aeronaut Alberto Santos-Dumont, whose first craft took to the skies above Paris in September 1898. 'My first impression,' he later recorded, 'was surprise to feel the wind in my face.' Santos-Dumont went on to develop a number of aerial runabouts and, in October 1901, he made the headlines by successfully flying his No.6 airship on a twelve-mile course across Paris and around the Eiffel Tower to claim the Deutsch Prize. Santos-Dumont's brave adventures continued to captivate the public's imagination, but his airships were tiny and in no way could fulfil Count Zeppelin's militaristic ambitions.

Last week I flew in a Zeppelin. We just cruised for forty minutes, but could open the windows, speak without effort, enjoy watching the world go by 1,000ft below and tell ourselves what it must have been like when far bigger airships were having their heyday … Airship flying is total joy from beginning to end and inbetween.

Anthony Smith, broadcaster and author, on flying in the Zeppelin NT, 2007

In contrast to all previous designs, Count Zeppelin's concept was to string a number of gas balloons together within a skeleton, consisting of lightweight metal rings and longitudinal girders, wrapped within a protective outer cover. The advantage of this type of 'rigid' airship is that the gas cells are able to expand or contract as required, in response to fluctuations in atmospheric pressure and temperature, without the risk of compromising the airship's shape.

Construction of the prototype Zeppelin, LZ1, began in 1898 and was completed inside a floating wooden hangar just off the shores of Lake Constance, in southern Germany. The reason for placing the hangar on the water was that it could rotate freely, allowing the airship to be removed in line with the wind direction. Problems with the hangar breaking its moorings, however, caused a delay of several months, but by 2 July 1900 the LZ1 was ready for her maiden flight.

The airship, designed by the brilliant young engineer Ludwig Dürr, who went on to design all of the subsequent Zeppelins, was a

Did you know?
In addition to carrying fare-paying passengers, the *Hindenburg* earned revenue by taking airmail and freight across the Atlantic. This included cars, a complete aeroplane plus a variety of animals being transported to zoos.

▶ *Civic pride in Count Zeppelin's work is marked by this fountain near the shoreline of Lake Constance at Friedrichshafen. It features the figure of a boy holding the first Zeppelin aloft.*

simple elongated cylinder with pointed ends. Its aluminium-zinc alloy frame was 420ft (676m) long and it contained seventeen rubberised-fabric gas cells with a combined capacity of 399,000cu ft (cubic feet) or 11,290cu m (cubic metres). Suspended beneath its underbelly were two open gondolas – boat-shaped cars – each with a 14.2hp Daimler engine driving aluminium airscrews, located on either side of the airship, via transmission shafts. It was equipped with rudimentary but largely ineffectual control surfaces, and a sliding counterbalance weight was suspended beneath the length of the airship to control its angle of attack or pitch. It proved to be a short maiden flight, lasting only eighteen minutes, and a disappointing one. In all likelihood the LZ1 had most probably drifted with the light winds on its 3.5-mile (5.5km) sojourn because of problems with the controls.

Some of the shortcomings of the LZ1 were soon rectified. The structure was strengthened through the addition of a triangular external keel, the Daimler Company was persuaded to lend two of its latest 80hp engines, and the moveable weight system was discarded in favour of box-kite elevator surfaces fore and aft. All of these improvements resulted in a creditable airship, but unfortunately the second flight of the LZ2, on 17 January 1906, ended in disaster. The rudder jammed and there were problems with both engines, leaving the airship drifting out of control. It eventually landed as a free balloon, only for the framework to be smashed by stronger winds during the night.

◄ The LZ6 being eased backwards out of the floating hangar on Lake Constance, 1909.

Once again Count Zeppelin picked up the pieces and proceeded with LZ3, based on her predecessor but with slightly upgraded engines and additional horizontal tail fins. Initial trials were encouraging, including one flight of eight hours' duration, and the German government began to take an interest. They offered to buy the airship and its successor if Zeppelin could achieve a non-stop demonstration flight of twenty-four hours to cover a minimum distance of 435 miles (700km). As LZ3 was not up to the task, work began on the LZ4 in March 1908, and this airship was to play a pivotal role in the future of the Zeppelin Company.

The LZ4 was larger at 446ft (136m) long, with a capacity of 529,720cu ft (15,000cu m), and, most importantly, she had two 105hp four-cylinder Daimler engines and larger control surfaces. She was an immediate success and, after a few preliminary test flights, including a twelve hour trip over Switzerland, on 4 August 1908 Count Zeppelin set off from Friedrichshafen at the start of the government-proving flight. When an engine breakdown forced an unscheduled landing near the small town Echterdingen, not far from Stuttgart, the ship was secured in the open while the repairs were made. During the night a sudden squall flung the LZ4 into the air, threw her into trees and the leaking hydrogen burst into flames, completely destroying the airship.

This time the seventy-year-old count believed that his dream was over. But

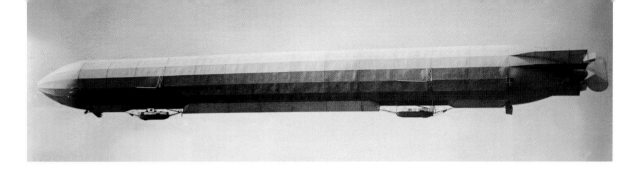

▲ LZ12 became the German army's ZIII. Launched in 1912, it was dismantled only two years later. (US Library of Congress)

the German people had other ideas and, in what became known as the 'miracle of Echterdingen', there was a massive and spontaneous groundswell of popular support and donations began to pour in. As a result the Zeppelin Company was formed, with new facilities in the town of Friedrichahfen, and the business of building airships carried on. Under the guidance of a new business manager, Alfred Colesman, a new company known as DELAG – the *Deutsche Luftschiffahrts Aktien Gesellschaft* (German Airship Share or Holding Company) – was formed in 1909 with the intention of linking major cities within Germany through scheduled airship flights.

In 1910 the LZ7 was christened as *Deutschland*, but met a swift end when she crashed in severe winds. This marked the start of another patch of bad luck for the Zeppelins: LZ6 burned in its shed and LZ8 was wrapped around its hangar doors by strong crosswinds. In command of the LZ8 on that occasion was Hugo Eckener, a former journalist who had been won

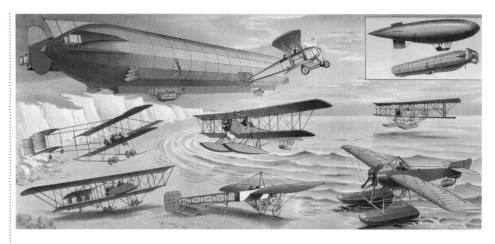

➤ *This illustration from c.1915 portrays the family of contemporary flying machines with the Zeppelin dominating the scene, plus biplanes, monoplanes, seaplanes and a non-rigid and another rigid inset.*

over to Zeppelin's cause and would later become its driving force in the interwar years. By the time of the LZ10 *Schwaben*'s first flight in 1911, DELAG was ready to embark on a brief period of regular passenger services. Over the next two years more ships followed: *Viktoria Luise* (named after the Kaiser's daughter), *Hansa* and *Sachsen*. By the outbreak of war in 1914, this fleet had successfully conducted a total of 1,588 commercial flights between Frankfurt, Düsseldorf, Baden-Oos, Berlin-Johannisthal, Gotha, Hamburg, Dresden and Leipzig.

At midnight, the three Zeppelins reached the city of London from different directions. The English could darken the metropolis as much as they liked, but they couldn't conceal the Thames. They even placed false streetlights in Hyde Park...but the airship officers were not deceived; the course of the Thames betrayed the ruse.

Heinrich Mathy, commander of LZ31 during a bombing raid on London, 25 September 1916

Zeppelins were clawed down in flames from the skies over both land and sea by aeroplanes until they dare not come any more. The aeroplane was the means by which the Zeppelin menace was destroyed, and it was virtually the only means, apart from weather and their own weakness, by which Zeppelins were ever destroyed.

Winston S. Churchill, *The World Crisis*

Did you know?
Although the *Hindenburg* was slightly less than twice the length of the first Zeppelin, LZ1, it was almost twenty-one times bigger in volume.

Between 1914 and 1918 the First World War tore through Europe like a ragged scar. This was the conflict that introduced mechanised warfare and slaughter on an unprecedented scale. It was also the first war in which aviation played a significant role and for the inhabitants of London, and several other British towns, it was their introduction to the concept of total war. One in which the enemy would reign death and destruction upon the Home Front.

At the outbreak of hostilities in 1914, the Zeppelin Company was running its commercial airship services between major

▲ *Forward gondola of a German army Zeppelin of the First World War, complete with defensive machine gun positions.*

aeroplane for that matter, introduced a new aerial dimension to warfare, and initially the German High Command responded by deploying its airships in support of the army at the front through close-support bombing. This was an ill-conceived approach and four airships were soon lost to ground fire. In the long term the airships would show themselves better suited to the roles of aerial reconnaissance, in particular spying on shipping, and carrying out raids upon England.

Kaiser Wilhelm had opposed bombing England at first, especially London, because of his close blood-ties with the British royal family. But such niceties were not supported by members of his naval staff, in particular Commander Peter Strasser, the chief of the German Naval Airship Division, who advocated the use of any means to

cities within Germany and their DELAG fleet was immediately pressed into military service. The German navy and army had both been operating a handful of airships by this time, built by the Zeppelin Company and their rivals, Schütte-Lanz, but with very mixed results. However, the airship, and the

THE SUPER-ZEPPELIN IN ITS PRIDE
NEW FEATURES WHICH DID NOT SAVE THE GIANT'S SKIN

bring England to her knees. By early 1915 the Kaiser relented and allowed the air raids to commence.

Providing Strasser with the tools to do the job was a new generation of airships, ordered before the war and now entering service. An improved version of the pre-war L3 type was known as the M-class: 518ft (158m) long, with a volume of 793,518cu ft (22,470cu m), it was powered by up-rated 180hp Maybach engines producing a top speed of around 50mph (80km/h). This was the first airship with any real long-distance capability.

Bad weather caused Strasser to abandon the first raid against England on 13 January 1915, then six days later three naval Zeppelins battled against strong winds in a bid to attack Hull. Driven off course, they ended up dropping their load of

◄ *The anatomy of a 'Super Zeppelin' as published by* The Graphic *magazine in December 1916.*

Did you know?
The US Navy's large rigid airships ZRS4 *Akron* and ZRS5 *Macon* were fitted with secondary auxiliary control stations located in their lower tail fins.

➤ *'The Raider' is caught in the searchlights, depicted in a propaganda postcard. Usually this image would be just one in a sequence of photographs culminating with the Zeppelin's fiery destruction.*

➤➤ *The Zeppelin raids were meant to terrorise the British but instead they provided powerful ammunition for the recruitment posters. (US Library of Congress)*

THE RAIDER.
PUBLICATION SANCTIONED BY OFFICIAL PRESS BUREAU
Publishing Office
39. St. Andrew's Hill, E.C.   (Copyright)

IT IS FAR BETTER
TO FACE THE BULLETS
THAN TO BE KILLED
AT HOME BY A BOMB

JOIN THE ARMY AT ONCE
& HELP TO STOP AN AIR RAID

GOD SAVE THE KING

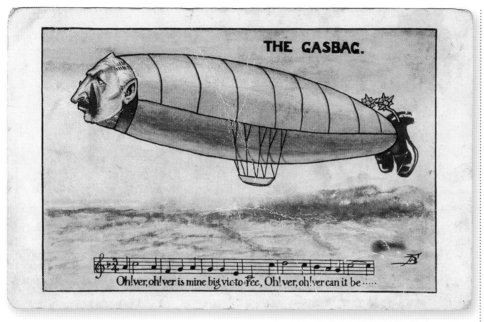

THE GASBAG.

Oh! ver, oh! ver is mine big vic-to-ree, Oh! ver, oh! ver can it be .....

twenty-four 110lb (50kg) bombs on Great Yarmouth and the Kings Lynn area of Norfolk. Four civilians died that night and another sixteen were injured. Further raids followed over the ensuing months, but the weather conditions minimised their

effectiveness and on several occasions the airships were forced back to base.

By the spring of 1915 the Kaiser reluctantly gave the go-ahead for bombing raids on London itself, principally on the docks. The bigger P-class Zeppelins had become available and these featured a 531ft (162m) hull constructed for the first time with a duralumin alloy to replace aluminium. This increase in size was combined with greater power, from an additional fourth engine, to create an airship that was around 10mph (16km/h) faster, and had an operational ceiling of 10,000ft (3,050m). The newer Zeppelins also incorporated the simplified cruciform tail and greater streamlining which had been demonstrated by Professor Schütte on his wooden-framed rigid airships.

On the evening of 31 May 1915, two airships of the Army Airship Service, LZ37 and LZ38, set off for London. The LZ37 was forced to turn back following damage to her outer cover, but LZ38 had no problem following the course of the Thames all the way to the capital. The airship's crew dropped incendiary bombs on a number of locations, causing several fires and resulting in four fatalities, including two children. Even though the Zeppelin raids had been widely expected, Londoners were shocked by the brutal reality of all-out war and the inability of the authorities to tackle the intruders. The LZ38 had passed almost unseen, unhindered by searchlights or hostile fire. Fifteen aircraft of the RNAS had taken off in pursuit, but only one pilot caught so much as a glimpse of the airship.

Further raids followed and while the drone of the airships' engines did cause fear among the civilian population, the attacks failed to create the widespread

▲ *The Zeppelin 'cloud car' was used to lower an observer beneath the clouds. This example is on display at the Imperial War Museum in London.*

► *Count Zeppelin: 'Stands London where it did my child?' The Child: 'Yes Father, missed it again.'* Punch *cartoon from 1915.*

►► *Recruiting poster celebrating Sub-Lieutenant Warneford's achievement in bringing down the LZ37 in June 1916.*

terror anticipated by the Germans – partly because the British Government had imposed restrictions on reporting the raids. More significantly, the Zeppelins provided the authorities with a powerful propaganda weapon and the nocturnal raiders were soon labelled as 'baby killers'.

Then began a game of technological catch-up. Gradually the British improved their defences with searchlights and anti-aircraft batteries surrounding London, as well as new aircraft which could climb higher and attack the Zeppelins with incendiary bullets to ignite their hydrogen gas. But by the spring of 1916 the first of the 'Super Zeppelins' had arrived on the scene, beginning with L30, and these could fly even higher while carrying 5 tons of bombs.

It wasn't until 7 June 1916 that the British claimed their first kill. Sub-Lieutenant Reginald Warneford brought down LZ37, which was returning across the North Sea after an aborted raid. Just over a year later, Second Lieutenant William Reefe

Robinson shot down SL-11 over Cuffley, Hertfordshire. Although this was actually a Schütte-Lanz airship, its destruction was celebrated as the first Zeppelin destroyed over British soil. These successes pushed the Zeppelins ever higher and, accordingly, each new type was bigger than any before, culminating with the L70-class Type-X of 2,196,300cu ft (62,200cu m), which had a theoretical range of around 7,460 miles (12,000km) and a ceiling of 20,000ft (6,000m).

In retrospect, the Zeppelin raids on England were an expensive exercise, both in terms of resources and the loss of airmen, and one which had little impact upon the outcome of the war. Even so, this period of unparalleled aeronautical progress would pave the way for the ensuing age of international air travel.

As the airships sailed along they smashed up the city as a child will shatter its cities of bricks and cards... Lower New York was soon a furnace of crimson flames, from which there was no escape.

H.G. Wells, *The War in the Air*, 1908

When flying at night, possibly on account of the darkness, there is always the feeling of utter loneliness directly one loses sight of the ground. We feel this loneliness very much tonight; possibly owing to the fact that we are bound for a totally unknown destination across the Atlantic.

Air Commodore E.M. Maitland, aboard the R34 on the first Atlantic crossing

The First World War saw the Zeppelins progress from simple craft that could travel between cities within Germany, to streamlined leviathans capable of linking the continents. This feat was most clearly illustrated by the flight of the L59 which, in November 1917, had departed from the southernmost Zeppelin base at Yambol, in Bulgaria, on an audacious rescue flight to take vital supplies to the beleaguered troops defending Germany's last African colony, Deutsch-Ostafrika, otherwise known as German East Africa. The intention was to make the 3,600-mile (5,800km) journey as a one-way trip. The airship was loaded with ammunition, machine guns, rifles and medical

supplies. Once it reached its destination the airship was to be cannibalised; the framework to be used to build barracks, the fabric and gas cells converted into sleeping bags or clothing, and the Maybach engines would run generators.

▲ *Hugo Eckener became the driving force behind the Zeppelins during the interwar years.*

➤ *The LZ126 shortly after arrival at the US Navy's airship base at Lakehurst, New Jersey, in October 1924. It is seen here before the application of its naval markings and new name Los Angeles.*

Under the command of Lieutenant-Commander Ludwig Bockholt, the L59 had already reached the Nile Valley in Egypt when a radio message came from Berlin recalling the airship, as the plight of the German forces had not been as bad as intelligence had originally suggested. Almost four days after it had taken off, the L59 and its weary crew touched down at Yambol once again. Flying non-stop in difficult conditions with huge daily extremes in temperature, they had covered

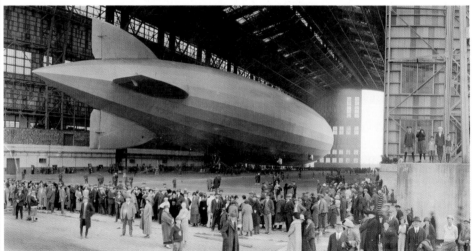

a distance of 4,225 miles (6,800km) with fuel to spare.

Clearly the flight of the L59 had demonstrated that a transatlantic crossing was achievable. Even before the war several unsuccessful attempts had been made to fly the Atlantic with smaller airships, most notably with Walter Wellman's *America* in 1910, which ditched after covering 900 miles (1,448km), but none had been successful. At the end of the hostilities the draconian terms of the armistice imposed

severe restrictions on Germany's ability to build new airships, and most of the existing Zeppelins were either scuttled in their hangars by their despondent crews or handed over to the various Allied countries as war reparations over the ensuing years. Despite the efforts of Hugo Eckener and others to generate interest in a transatlantic airship service, it began to look as if the Zeppelin Company might be forced out of business altogether.

The British had constructed several rigid-framed airships of their own and were greatly influenced by the design of wartime Zeppelins, which had literally fallen into their hands. The R34 was built by the Beardmore Company at Inchinnan, near Glasgow, and was the second of two R33-class airships modelled on the German L33, which had come down largely intact near Little Wigborough in Essex. Completed too late for its intended maritime patrol duties, the Air Ministry decided instead to send the R34 to the USA and back as a demonstration of the airship's long-distance capabilities.

Major George Herbert Scott was in command of a crew of thirty when the R34 departed from the airship station at East Fortune in the early hours of 2 July 1919. In the darkness the airship set course north-west for the Clyde and the Atlantic Ocean beyond. Accommodation aboard the airship was very basic as it had never been designed with passenger comfort in mind, and between watches the crew took turns to share hammocks slung from the girders along the length of the internal keel. For much of the flight, fog and cloud obscured the view as the R34 ploughed on at the relatively low altitude of 2,000ft (600m) or less, and at a reasonable speed of around 66mph (106km/h), thanks to a favourable tail wind. In many ways this flight was a test of the airship's potential for commercial transatlantic flights and the smoothness of the ride compared very favourably with surface vessels. For anyone prone to seasickness, the airship was the

Did you know?
The first female stewardess on a Zeppelin was Emile Imhof, who joined the crew of the *Hindenburg* in 1937.

▶ *The moment of the R34's departure from East Fortune, 2 July 1919, at the start of the transatlantic flight.*

way to travel. With little to see, the greatest excitement came with the discovery of a stowaway. William Ballantine had been bumped from the original crew to make way for more important guests of the Air Ministry. Luckily for him, he was not discovered until they were already out over the Atlantic or otherwise he would have been dropped off by parachute.

On 4 July land was spotted. It was the northern coastline of Newfoundland. By now the R34 was driving against strong headwinds and didn't reach her landing site at Mineola Field, on Long Island, until 9.54 a.m. local time on the morning of 6 July. The transatlantic crossing had taken just over 108 hours, establishing a new endurance record for airships. Thousands of New Yorkers flocked to the airfield to see this modern marvel, and for the crew there were a few days of rest and the chance of a hot bath before the return journey three days later. Clearly, this successful two-way transatlantic crossing was an emphatic vindication of the airship's long-distance credentials.

The frustration of Hugo Eckener in Germany can only be imagined as he watched his upstart rivals claim the transatlantic crown, but Eckener had greater concerns about the very survival of the Zeppelin Company. The Allies had imposed restrictions on the size of airships he could build, lest they be used for military purposes. Desperate to prevent the break-up of his team of highly-skilled workers, Eckener planned to restart the DELAG operations with two smaller airships. The 796,350cu ft (22,550cu m) LZ120 *Bodensee* and the slightly larger

▲ The forward section of the R33's control car, on display at the RAF Museum Hendon, London.

LZ121 *Nordstern* were completed in 1919 and had actually started flights between Friedrichshafen and Berlin when the Inter-Allied Control Commission snatched them in lieu of the wartime Zeppelins which had

➤ *The LZ126 about to enter the hangar at Lakehurst.*

been deliberately wrecked by their crews. To make matters worse, the American president, Woodrow Wilson, demanded that Zeppelin should build a brand new 2,500,000cu ft (70,750cu m) airship to be handed over to the USA. Fortunately, Eckener saw in this situation the opportunity to build the most advanced airship to date.

Work began on the LZ126 in 1922 and the completed airship made its maiden flight on 27 August 1924. Known to the Germans as the *Amerikaschiff*, or 'America Ship', when she landed at the Naval Air Station at Lakehurst, New Jersey, on 15 October 1924, she became the property of the US Navy and was christened the *Los Angeles*. Not only had Eckener ensured the continuation of the Zeppelin Company by proving beyond doubt that they were still the best in the business, he was also taken aback by the rapturous welcome they had received from the Americans. In more senses than one, a bridge had been built across the Atlantic.

It gives me and the people of the United States great pleasure that the friendly relations between Germany and America are reaffirmed, and that this giant airship has so happily introduced the first direct air connection between the two nations.

President Coolidge, upon the arrival of the LZ126 in the USA, 1924

With Germany's aviation industry on hold in the years immediately after the First World War, the way was left open for its former rivals to push forward with their own aerial ambitions. Great Britain, in particular, with its far-flung territories, saw itself at the hub of aeronautical development and, following the successful Atlantic flights of the R34, the airship was seen by many as the solution for long-distance transportation to link the empire.

The Imperial Airship Scheme, put forward by Sir Dennistoun Burney of the airship building company Vickers, was taken up by the newly elected Labour government in 1924. Initially, the plan was to construct two rigid-framed airships of 5 million cu ft (141,500cu m) volume and each capable of carrying 100 passengers a distance of up to 3,500 miles (5,600km). One of these prototype airships would be government-built at the Royal Airship Works at Cardington, Bedford, while the other was to be privately built by the Airship Guarantee Company, a subsidiary of Vickers.

In design, the two airships showed a deliberate move away from the German Zeppelins, with a more swollen 'fat cigar'

▲ The twin sheds at
Cardington still stand
as gigantic monuments
to the Imperial Airship
Scheme to link the
empire. Photographed
from the Zeppelin NT07
003 in August 2008.

long and with a maximum girth of 133ft (40.5m), its actual volume was slightly over target at 5,156,000cu ft (146,000cu m). Passenger accommodation for both airships was located within their hulls which meant that it was far roomier than any other airships so far. This included spacious saloons, observation decks and comfortable cabins. It is interesting to note that the R101 even had an asbestos-lined smoking room on the lower deck, thus preceding the *Hindenburg* by several years. One difference between the two British ships was the choice of engines. The R100 was fitted with six Rolls Royce Condor petrol engines, two of which had reversing capabilities for improved mooring manoeuvring, while the R101 had five Beardsmore Tornado diesels, including a rear engine that faced backwards.

profile. The R100, in particular, which was designed by Vickers' engineer Barnes Wallis and built at Howden in Yorkshire, featured only thirteen longitudinal girders giving it a flat-sided appearance. At 709ft (216m)

In the event, it was the R101 that won the race to be first in the air, on 14 October 1929. The R100 was two months behind, launching on 16 December. Following a number of test flights, including a fifty-four hour endurance flight over south-west England and the Channel Islands, the R100 was prepared for a proving flight to Canada and back. On 29 July 1930 the airship departed from Cardington on the 3,870-mile (6,280km) flight to land at the Saint-Hubert Airfield, near Montreal, almost seventy-nine hours later. The transatlantic crossing hadn't been without incident, however, and, driving against strong winds over the Strait of Belle Isle, they encountered violent squall conditions which rolled the airship and caused rips in the fabric on the lower and starboard fins. One hole was described by a crew member

▲ *Designed by Barnes Wallis, the 'capitalist airship' R100 riding the high mast at Cardington.*

as being 'large enough to drive a double-decker bus through'. Later on, a strong up-draught abruptly tossed the airship's nose skywards and pitched it down again, sending every moveable object tumbling and causing further damage.

Having reached Canada, the crew of the R100 spent twelve days in Montreal while the airship was repaired, refuelled and stocked for the return flight, which

➤ *Cutaway illustration showing the passenger accommodation located within the hull of the R100.*

First Gasbag

Double Staircase

Dining Saloon

Balcony 4 feet wide

Service Hatch and Box

Pantry

Service Lift

Promenade 40 feet long 6 feet wide

Lower tier Passenger Cabins

Provisions

Girders forming girder joints

➤➤ *The R101 at Cardington. The circular holes are 'gills' for ventilation; the entry hatch can be seen in the lowered position beneath the nose.*

began on 13 August 1930. Blessed with a tail wind, the R100's homeward journey took just under fifty-eight hours. Back at Cardington, the R100 was returned to her shed while every effort was directed to getting the R101 ready for an even more ambitious flight, to take the Air Minister, Lord Thompson, to the Imperial Conference being held in India in October 1930. Found to be lacking in sufficient lift initially, during the summer the R101

was chopped in two and extended with an extra gas bay, increasing her length to 777ft (237m) and volume to 5 million cu ft (160,000cu m). Under pressure from Thompson, the enlarged R101 was rushed through a series of woefully inadequate test

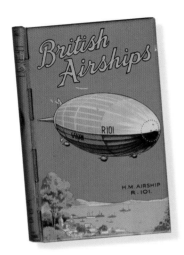

▲ *This novelty money box produced by Chad Valley reflects the nation's pride in the R101. It was the Concorde of its age.*

➤ *The R100 at Saint Hubert's airfield, near Montreal, after crossing the Atlantic in 1930.*

flights, and on the evening of 4 October 1930 she detached from the mooring mast for the first leg of the India flight, with a stopover scheduled at Ismailia, Egypt.

From the start many onlookers observed that the airship was struggling to gain height, and around 4 tons of water ballast was dropped before she disappeared into

the overcast. On board was a crew of forty-two, plus fourteen VIPs, including Lord Thompson. After supper most of them turned in for the night, but the airship was already encountering blustery headwinds over France which made her roll, possibly causing a loss of precious hydrogen. The R101 was making little headway – thought to be only 20mph (30km/h) or so – when she suddenly lunged downwards, momentarily righted herself and then dived again. The nose of the R101 struck rising

R101 dropping water ballast as she backs away from the mooring mast at Cardington.

ground near the small town of Beauvais and within seconds the entire airship was engulfed in flames as the hydrogen burned. Forty-eight men lost their lives, while only eight managed to clamber clear of the inferno, and of those two more died of their injuries over the next few days.

In that terrible moment Britain lost the cream of its airship men, and the future of the Imperial Airship Service was thrown into uncertainty. The R100 remained in her shed while the politicians debated her fate. She never flew again and by November the following year, 1931, work began on dismantling her framework, which was sold off as scrap metal.

The *Graf Zeppelin* is more than just machinery, canvas and aluminium. It has a soul…

Lady Grace Drummond-Hay, 1929

I have always felt that such effects as were produced by the Zeppelin airship were traceable to a large degree to aesthetic feelings. The mass of the mighty airship hull, which seemed matched by its lightness and grace, and whose beauty of form was modulated in delicate shades of colour, never failed to make a strong impression on people's minds.

Dr Hugo Eckener

Having kept the Zeppelin dream alive through the construction and delivery flight of the LZ126 – the *Amerikaschiff* – Hugo Eckener was rewarded by a relaxation of the post-war restrictions imposed by the Allied Control Commission. With an eye on the transatlantic passenger service, the LZ127, *Graf Zeppelin* was to reassert Germany's lighter-than-air lead and, in the process, became the most famous airship in the world.

In comparison with the later Zeppelins, in particular the *Hindenburg*, the profile of the LZ127 is much slimmer, more like a thin pencil, with a length-to-diameter ratio of 7.8 to 1. In fact, the airship was smaller than Eckener would have liked as the limiting factor was the size of the

from the government. The overall length of the LZ127 was 775ft (237m), with a maximum diameter of 100ft (30.5m). Her volume of 3,707,550cu ft (105,000cu m) made her the largest airship built at that time (although not as big as the British R100 and R101 which were not completed until after the LZ127).

Within the duralumin framework, the hull was divided into two spaces; the upper two thirds were occupied by gas cells for the hydrogen and the remaining lower space was fitted with cells to contain a gaseous mixture of propylene, methane, butylene and hydrogen. Known as 'Blaugas', this cocktail of gases was only very slightly heavier than air and would be consumed as fuel by the five 550hp Maybach VL2 engines without significantly altering the airship's buoyancy. The consumption of

▲ *Pre-dawn and the nose of LZ127* Graf Zeppelin *is seen behind the hangar doors at Friedrichshafen.*

construction hangar at Friedrichshafen. The Zeppelin Company could not afford to build new facilities and only raised the money to build the *Graf Zeppelin* through a public fund and, later, with a contribution

◄ *Construction workers assemble one of the main rings of the LZ127* Graf Zeppelin.

▲ *The* Graf Zeppelin *flying over Friedrichshafen, the historic home town of the Zeppelin Company.*

wide, located near the front of the airship. This included the control room which housed the rudder and elevator wheels, gas valves and ballast controls, and behind that the map and radio rooms. On the starboard side was a compact electric-powered galley or kitchen, and then came the lounge or saloon, with a corridor leading to ten double-berth cabins for the passengers. Each cabin featured a couch which folded upwards to form a bunk bed, a small table and an external window. The toilets and washrooms were at the rear of the gondola.

Passengers were expected to spend most of the time in the saloon area which also doubled up as the dining room. The accommodation was not as comfortable as a transatlantic ocean liner perhaps, the menu maybe not as lavish, but this was more than made up for by the faster

conventional fuels causes an airship to become progressively lighter on long flights as the fuel is used up.

Compared with the enormous size of the hull, the accommodation was not especially roomy. It was arranged within an external gondola, 98.5ft (30m) long and 20ft (6m)

crossing times. Many passengers also appreciated the absence of seasickness when travelling by airship.

The *Graf Zeppelin* first took to the air on 18 September 1928, and such was the confidence in the new airship that after only five test flights she departed on the inaugural transatlantic run to New York less than a month later, on 11 October 1928. Financing for the flight had come from a number of sources including the carriage of special airmail and postal covers, fare-paying passengers and a contingent of newspaper reporters and cameramen. Among them was the British journalist Lady Grace Drummond Hay who, along

◄ ▲ *Control cabin of the* Graf Zeppelin *with the rudder-man's position facing forwards at the front and the elevator controls on the left-hand side.*

▲ *Side view of the* Graf Zeppelin*'s control cabin, showing the cushion beneath the gondola.*

▲ *The stylish dining room of the* Graf Zeppelin *also doubled up as the public lounge or saloon area.*

and the catering arrangements. It proved to be something of a picnic lifestyle as she confided: 'The tiny kitchen is inadequate to supply luxuries for passengers as well as cook innumerable cans of steaming food for the thirty-nine members of the crew.'

Taking charge on the first crossing, and on many of the subsequent flights, was Hugo Eckener himself. Widely recognised as the most skilful of airship commanders with a legendary and almost intuitive grasp of the weather's every nuance, Eckener was also an affable host who inspired great confidence in his crew and the passengers. This was just as well, for although he had selected a southerly route to avoid the seasonal storms expected to the north, the airship was caught in a powerful squall line out over the Atlantic. The nose pitched downwards and in response the crewman

with her colleague Karl H. von Weigand, represented Hearst's Newspapers, and their day-by-day accounts of the voyage were devoured by a news-hungry public. Lady Drummond Hay described the comforts of airship travel, including the tiny cabins

◄ *The artist Ludwig Dettmann joined official photographers aboard the* Graf *on the first transatlantic flight. This is his pastel impression of the airship encountering stormy weather above rough seas.*

◄◄ *Passenger cabins aboard the* Graf Zeppelin *were very compact. The rear of the couch folds upwards to form a bunk bed.*

on the elevator controls over-reacted, thrusting the airship's nose upwards. It was breakfast time in the saloon and the passengers found themselves, the furniture and their breakfast, tumbling into a heap at one end. The lone voice of calm in the confusion was Lady Drummond Hay, who laughed out loud at their predicament.

The *Graf* withstood her violent shake-up, although a large part of the outer cover had

been ripped from the port side fin. Eckener slowed her down to half-speed, radioed the US Navy to have a vessel on standby, and dispatched a repair team to the damaged fin. Among them was Eckener's son, Knut. It was a race against time as the airship slowly settled towards the grey waters. At a height of around 300ft (90m), Hugo Eckener had no choice but to increase engine speed and, thankfully, the men working on the tail had managed to scramble to safety in the nick of time. It was a close call which added to the drama and excitement of this first scheduled passenger crossing of the Atlantic by airship and served to heighten the enthusiastic reception they received upon their arrival at the US Navy's airship base at Lakehurst, New York. The flight from Friedrichshafen in Germany had taken a little under 112 hours, and on 28 October 1928 the *Graf Zeppelin* made the return trip in under seventy-two hours.

DEUTSCHE ZEPPELIN-REEDEREI

◄◄ *If ever a city cried out to be seen from a Zeppelin, then it must be New York.*

◄ *An advertising brochure for the* Graf Zeppelin's *transatlantic passenger service.*

55

At the beginning, it is hard to realize you are on board a Zeppelin; the comfort and protection from the weather, the spaciousness, the elegance and neat equipment of the well-appointed cabins, the deference of the ship's company who are only too ready to help, awake in you a new conception of pleasurable travel.

Airship Voyages Made Easy, Deutsche Zeppelin Reederei

Man-made air monsters, as big as our greatest skyscrapers, will ride the ocean skyways like mighty silver-plated hotels magically floating on high. London will be within two dawns of Broadway… We shall skim the icy rim of the world in weekend sightseeing jaunts to the North Pole and roar around the earth in a week.

'Two Days to Europe in a Flying Hotel', *The American Magazine*, May 1930

Following on from the triumphant first transatlantic roundtrip of the LZ127 *Graf Zeppelin*, the second westward hop across the pond almost ended in disaster before it had barely begun. The airship departed from Friedrichshafen on the morning of 6 May 1929 and was making headway towards the Mediterranean, aided by the

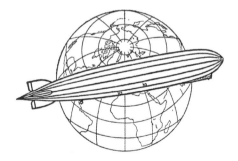

Mistral winds, when one of the engines failed. In itself this was no great problem as the *Graf* often flew with only four engines to save on fuel. Eckener was pushing on towards Gibraltar when a second engine gave up. With only three out of five engines operational, he knew that they had to return to Friedrichshafen, but by then

➤ *Route of the* Graf Zeppelin's *historic 1929 round-the-world flight.*

Did you know?

The ride on the big transatlantic Zeppelins was so smooth that a popular pastime with passengers was to balance a pencil on its end to see how long it remained upright without falling over.

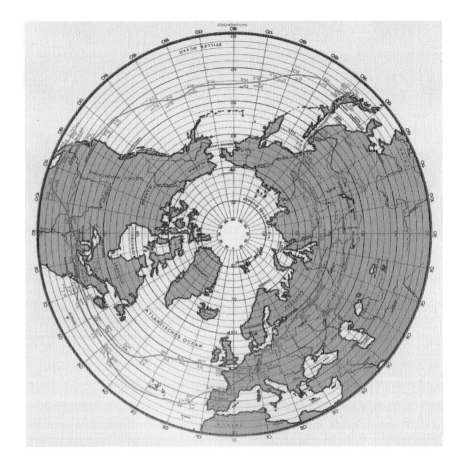

they were flying into the increasingly strong winds. The following morning they were still struggling to make headway when two of the remaining Maybach engines stopped. Eventually Eckener made a landing at the airship base of Cuers-Pierrefeu, in southern France, to wait for replacement engines from Germany.

For some commentators this incident was an indication of the airship's vulnerability, but to Eckener it merely confirmed that airships could safely overcome such difficulties in flight. To prove the point, the next voyage of the *Graf Zeppelin* was to be the most ambitious and remarkable airship journey ever undertaken: a voyage around the world. Because the financing for this flight was coming from the wealthy American newspaper publisher William Randolf Hearst, Eckener was required to

◄ *This German publication shows the arrival of the* Graf Zeppelin *in Tokyo after the longest leg of the round-the-world flight.*

The exploits of the German-built Graf Zeppelin celebrated on an American postage stamp.

Following the successful conclusion of the round-the-world flight, Hugo Eckener and his crew were celebrated as modern-day heroes by the New Yorkers.

begin and end the circumnavigation in the USA. Accordingly, on 1 August 1929, the *Graf* headed to Lakehurst, New Jersey, before officially beginning the first leg of the global flight which took her straight back to Germany.

On board were a number of specially invited guests and a contingent of reporters and cameramen. Lady Drummond Hay and Karl von Weigand would provide reports for Hearst's newspapers, as they had done on the first transatlantic crossing the previous year. At Friedrichshafen the airship was refuelled for the longest part of the flight, the 7,000-mile (11,260km) leg across the Siberian wilderness to Japan. When she landed at the Kasumigaura naval air station in Tokyo, the *Graf*'s passengers and crew received a rapturous welcome. They departed three days later to make the first non-stop transpacific crossing, landing at Mines Field in Los Angeles before proceeding to Lakehurst. This had been the first round-the-world flight by any type of aircraft and, in the process, the *Graf Zeppelin* had clocked up 19,500 miles (31,400km) from start to finish – even more if you count the initial flight from Germany to the USA. It had been a phenomenal achievement which served to

▲ *In 1931 the* Graf Zeppelin *flew north to the Arctic Circle for a rendezvous with the Russian ice-breaker Malygin.*

a cruise over southern Europe and into Palestine in the spring of 1929, and the following year she embarked on a polar expedition to rendezvous with a Soviet icebreaker deep within the Arctic Circle. In May 1930 the *Graf* made her inaugural trip over the South Atlantic to mark the start of regular passenger services between Germany and Brazil.

The *Graf* wasn't the only airship to make its mark on the international stage. As we have already seen, Britain's R34 had shown the way across the Atlantic back in 1919, the LZ126 had made a one-way crossing in 1924, and in 1930 the R100 completed the double by flying to Canada and back. There were also the two Italian-built airships, N-1 *Norge* and N-4 *Italia*, which flew from Europe to North America over the polar ice cap. Unlike the rigid giants, the N-1

establish the airship's apparent dominance of long-distance travel at a time long before passenger aircraft began operating across the oceans.

There were other headline-grabbing flights for the *Graf Zeppelin*, including and N-4 are termed as semi-rigids as they featured pressurised envelopes arranged above a rigid keel – a design favoured by the Italian airship pioneer General Umberto Nobile. In 1925 Nobile had been contacted by the Norwegian explorer Roald Amundsen, who planned to make the first flight to the North Pole. The Italian airship N-1, renamed as *Norge* (Norway), was of medium size with a volume of 671,000cu ft (19,000cu m) and a length of 348ft (106m). As with the Zeppelins, she was powered by Maybach engines, three of them, attached to the rigid keel which also incorporated the control car, other accommodation and supplies.

The *Norge* set off from Rome on 29 March 1926, and flew via the UK and Oslo to Kings Bay on the island of Spitzbergen, which lay within the Arctic

NACH
SÜDAMERIKA
IN 3 TAGEN!

GENERALVERTRETUNG DES LUFTSCHIFFBAU ZEPPELIN
HAMBURG-AMERIKA LINIE

◄ Poster for the Hamburg–America Line. The Graf *was considered too small for the North Atlantic run to New York, and she mainly flew on the South Atlantic route between Germany and Brazil.*

Circle. On 11 May she departed with a team of seventeen on board, including

Amundson, Nobile and Lincoln Ellsworth of the USA. This international mix created a conflict of egos and when the flags of the three principal nations were dropped at the North Pole there was some argument about whose had been the biggest. All went smoothly otherwise, and by the third day the *Norge* had landed at the Eskimo village of Teller in North Alaska, where it was dismantled as no arrangements had been made to fly her back to Europe.

General Nobile felt that this voyage did not sufficiently reflect his country's role in the endeavour, and so he mounted a second polar expedition with the slightly larger N-4 *Italia* in 1928. Alas, this attempt ended in disaster when the she became weighed down by ice and crashed, leaving only nine of the men alive and stranded on the ice cap. In the ensuing rescue mission Amundsen was killed when his search aircraft crashed. The survivors of the *Italia*'s crew, including Nobile himself, were eventually rescued but instead of receiving a hero's welcome, Nobile returned to Italy in disgrace and he later moved to the Soviet Union to continue his work with semi-rigid airships.

▲ *Italy's leading airship figure was General Umberto Nobile, designer of the N-1* Norge *and N-4* Italia *semi-rigid airships.*

◄ *The N1* Norge *at Pulham, Norfolk, on its way north to the Arctic Circle for the first crossing of the North Pole from Europe to North America in 1926.*

◄◄ *The* Graf Zeppelin *at the landing field at Recife, near Rio de Janeiro, in 1934.*

I don't see how long-distance reconnaissance is going to be carried out without using dirigibles, and the rigid appears to be a better type for that than the non-rigid… There certainly does not seem to be any very great promise in aeroplanes for long-distance scouting. It would appear that you have to go into dirigibles for that purpose.

Captain E.J. King, reporting to the General Board of the US Navy, 1918

During the 1920s a number of schemes to create home-grown passenger airship services in the USA, many promoted in partnership with the Zeppelin Company, came to nought and the large rigid airship remained the sole province of the US Navy. The airship's proponents within the Navy saw them as airborne scouts for their fleets – an 'eye in the sky' – but with the exception of the LZ126 *Los Angeles* their involvement seemed blighted right from the start.

The first American rigid airship, the ZR1 *Shenandoah* (Daughter of the Stars), was basically a close copy of the Zeppelin L49, a wartime height-climber. With components built by the naval aircraft factory in Philadelphia, and assembled at Lakehurst, the ZR1 was slightly longer than the Zeppelin, giving it a greater gas volume of 2,151,200cu ft (60,915cu m). This additional volume was vital as the airship was to be filled with helium which is slightly less efficient as a lifting gas than hydrogen. The airship first flew on 4 September 1923, and photographs reveal a slender ship with obvious First World War ancestry, featuring

a control cabin slung below the hull. Over the next two years the ZR1 was flown extensively until, on 3 September 1925, she broke apart in mid-air after encountering a line squall while on the way to a state fair in Ohio. The airship snapped into three pieces and the nose section spun like a top as it

◄ *American recruiting poster featuring biplanes, an observation balloon and airship. (US Library of Congress)*

▼ *The US Navy's first rigid airship, the ZR1 Shenandoah,* was a copy of a wartime Zeppelin.

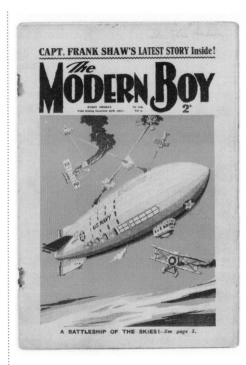

This 1931 magazine depicts a US Navy airship as 'a battleship of the skies', complete with attacking biplanes and guns blazing away.

CAPT. FRANK SHAW'S LATEST STORY Inside!

The MODERN BOY

A BATTLESHIP OF THE SKIES!—*See page 3.*

rose to 10,000ft (3,050m) and then slowly fell back to earth. In total twenty-nine men survived in the two larger sections.

Incredibly the ZR2 had already suffered a similar fate. Unable to get their hands on a German reparations airship, the Americans had turned to their British allies to build them one. The R38, which was the British designation for the ZR2, had been undergoing pre-delivery trials when she snapped in two over the Humber on 24 August 1924. It is thought that extreme rudder movements, designed to simulate the stresses likely to be encountered in severe weather over the Atlantic, caused the framework to fail. On board were seventeen Americans and thirty-two British crewmembers. Only five survived.

Thankfully, for the lighter-than-air faction within the US Navy, their next airship, the

Zeppelin LZ126, performed outstandingly well and proved to be the most long-lived of all the rigid airships. By 1926 confidence had returned sufficiently for the US government to give the go-ahead to build two massive airships, the ZRS4 and ZRS5, which would serve as flying aircraft carriers. Designed by the former Zeppelin engineer Dr Karl Arnstein, they were built by the Goodyear-Zeppelin Corporation, which had been formed as a joint venture between the two companies to allow for airship construction in the USA after the First World War. The identical airships would be 785ft (239m) long and have a volume of 6,850,000cu ft (193,970cu m). The largest helium-filled airships ever built, they were just

➤ *A Curtiss Sparrowhawk aircraft hoisted aboard the* USS Akron *after mating its hook in the trapeze mechanism, photographed in 1932. (US Navy)*

20ft (6m) shorter than the hydrogen-filled *Hindenburg*.

Inflated with inert helium, it was possible to locate the engines inside the hull, instead of within individual pods, and they drove propellers that could be swivelled to direct their thrust. Another innovation was the use of equipment to recover water from the engine exhaust gasses to be used as ballast to compensate for fuel burnt in flight. But perhaps the most noteworthy aspect of the design of both the ZRS4 and ZRS5 was the system for launching and retrieving aircraft in-flight from the airships' bellies. This was a concept pioneered by the British, and greatly refined with the US Navy rigids. Both airships had an internal hangar space which could accommodate up to four Curtiss F9C-2 Sparrowhawk biplanes fitted with skyhooks to their upper

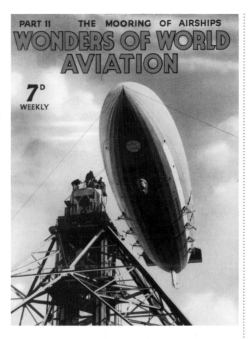

PART 11   THE MOORING OF AIRSHIPS
WONDERS OF WORLD AVIATION
7ᴰ
WEEKLY

◄ *The ZRS5 USS* Macon *– a wonder of world aviation.*

wings. On approaching the airship the pilot would push the skyhook through a looped trapeze hanging beneath the airship

71

➤ *The USS* Akron *flying over Manhattan, New York.*

and it could then be hoisted aboard. The advantage of this concept was to massively increase the footprint that the airship and her Sparrowhawks could patrol. In addition, the aircraft would have been deployed to defend the airship from attack.

Christened the USS *Akron*, the ZRS4 was launched in 1931 and was in service for two years before it was caught in bad weather off the Atlantic coast of New England. The tail hit the water and the airship broke up with the loss of seventy-three crew members. The USS *Akron*'s sistership, the almost identical ZRS4 USS *Macon*, was launched at Lakehurst just three weeks after the crash. In an extraordinary case of history seeming to repeat itself, on 12 February 1935 the *Macon* fell into the Pacific after wind shears caused a structural failure at the point where the upper fin was attached. It is likely that the framework had been weakened at this point in a previous incident. The USS *Macon* came down on to the water's surface relatively gently and as a result all but two of the crewmen were saved.

In 1991 the debris field of the USS *Macon* was located by the Monterey Bay Aquarium Research Institute, and subsequent investigations were carried out in association with the National Oceanic and Atmospheric Administration (NOAA) in 2006. Cameras fitted to remotely operated vehicles working at depths of 1,500ft (460m) have returned remarkable images of the airship's remains, including the Sparrowhawks. Now listed on the US National Register of Historic Places, the *Macon*'s final resting place is designated as a US Navy grave site.

Did you know?
The word hangar comes from the French. In the UK an airship hangar is known as a 'shed', while in Germany they are called '*halle*' or '*zeppelinhalle*'

➤ A remarkable image showing the wing of a Curtiss biplane, photographed by a remote-controlled submersible during investigations of the wreck of the USS Macon in 2006. (NOAA)

World traffic via airships has begun. By the expansion of weather-services on land and sea on the one hand, and the increased safety, comfort and practicality on the other, it will spread out over all the seas and continents. The world should be grateful to Germany as the trailblazer.

Ernst Lehmann on the conclusion of the *Hindenburg*'s first transatlantic season in 1936

Following on from the successes of the LZ127 *Graf Zeppelin*, Hugo Eckener proposed a new and expanded version, the LZ128, which would offer accommodation for thirty to thirty-four passengers. As with all previous Zeppelins, she was to be inflated with hydrogen but, following the loss of the R101 in October 1930, Eckener determined that the next generation of passenger airships should be flown with non-flammable helium. The LZ128 was passed by and in its place came the bigger LZ129, later christened the *Hindenburg*.

➤ *A new larger hangar was built at Friedrichshafen for the construction of the LZ129 Hindenburg.*

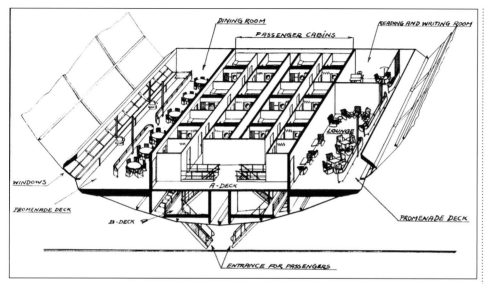

Diagram of the accommodation on board the Hindenburg. The cabins are located centrally, with the public areas to either side.

This was a truly awe-inspiring leviathan, the largest the world has ever seen – 805ft (245m) long with 7,062,900cu ft (199,880cu m) of lifting gas.

In keeping with standard Zeppelin practice, the LZ129 had a duralumin framework with fifteen main rings, giving a maximum diameter of 135ft (41.2m) –

almost twice that of the *Graf Zeppelin*. The number of main longitudinal girders running lengthwise was increased to thirty-six and, with intermediate girders in between, the result was a smooth streamlined hull. Unlike the *Graf*, there was no facility for

carrying Blaugas as fuel, and instead the sixteen gas cells would contain helium. At least that was the plan. Power would be supplied by four 1,200hp diesels developed by Maybach and later known as Daimler-Benz DB602 engines.

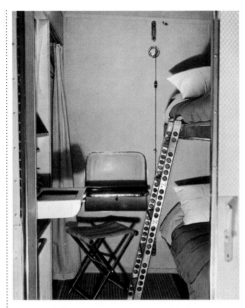

▶ *One of the Hindenburg's compact cabins complete with bunk beds and fold-up table and sink. These internal cabins had no windows, although this was rectified to some extent after the addition of further cabins for the 1937 season.*

In 1930 work began on the LZ129, but it soon became clear that the Americans, who held the world's main sources of helium gas, were not going to let Germany have any as they were fearful that the Nazi regime might deploy the airship for military purposes. Reluctantly, Eckener approved the modifications necessary to inflate the LZ129 with hydrogen and on 4 March 1936 she emerged from the Friedrichshafen hangar to make her maiden flight. The hull bore no name on that three-hour flight, and it has often been suggested that this was because Eckener was being pressured to name it after the Führer. But in truth the name *Hindenburg* had already been chosen and it was emblazoned in red gothic-style lettering by the sixth test flight. Nevertheless, the Zeppelin Company was not immune to the powerful political forces within Germany. The Air Ministry had provided the funding to complete the airship and by 1935 a new operating company, known as the Deutsche Zeppelin Reederei (DZR), was formed with the involvement of the state-owned Lufthansa airline.

If the mechanics of the LZ129 were impressive, then the accommodation was beyond comparison with any other form of air travel. Travelling on the new airship has been described as similar in comfort and elegance to an ocean liner. Passengers entered the underbelly of the airship via a pair of aluminium stairways. Inside, the accommodation was arranged on two decks. On the upper A deck there were twenty-five double birth cabins with large public spaces, dining areas, lounges, plus promenade decks arranged on either side of the hull. The cabins were compact, but functional enough with bunk beds and a foldaway sink with hot and cold water.

'The saloons provide ample lounging space for the fifty passengers which the ship is designed to carry,' recalled one passenger in 1936. 'The port saloon contains the dining accommodation, cleverly isolated by modernistic metal railing, the starboard holds a music room and small writing room. A feature of the former is a very handsome grand piano.' This special Blüther baby grand, weighing only 400lb (180kg), was constructed in

◀ *At the Zeppelin Museum in Friedrichshafen a full-scale section of the* Hindenburg *has been constructed, including this faithful re-creation of the lounge and promenade areas.*

aluminium and covered with yellow pigskin.

On the lower B deck there was the kitchen, a spacious affair compared with the cramped galley on the old *Graf*. It had an aluminium electric stove with four rings, plus roasting and baking ovens to make fresh bread, and a refrigerator. The remainder of B deck housed toilets, a shower room, crew facilities, the chief steward's cabin, plus a small bar and smoking room in which passengers could enjoy a cigarette or cigar under the strict supervision of the bar steward. This was entered via an airlock and the room was maintained at slightly higher air pressure to prevent the admission of any hydrogen.

Following a propaganda flight to drop election leaflets over German cities, the *Hindenburg* was made ready for her first transatlantic trip, not to the USA but to Rio de Janeiro. Hugo Eckener was on board, but his reluctance to kowtow to the Nazis had seen him relegated to the status of a 'non-person' and it was the former wartime

◄ *Side view of the 804ft-long LZ129* Hindenburg *at Lakehurst, bearing the Olympic rings to mark the 1936 Berlin Games.*

Zeppelin commander Ernst Lehmann who commanded the ship.

The *Hindenburg*'s first flight across the north Atlantic followed soon afterwards on 6 May 1936, landing at Lakehurst after an uneventful sixty-hour crossing. On returning to Germany she landed at the new international Zeppelin terminal at Frankfurt, and this would serve as the main base for future passenger operations. During that first season the *Hindenburg* made a total of seventeen round trips across the Atlantic; ten to Lakehurst and the remainder to Rio. She also made several internal flights at the request of Goebbel's Propaganda Ministry, including an appearance above the Berlin Olympic Games in August 1936. In addition, the *Graf Zeppelin* had been kept busy servicing the South Atlantic route, and justifiably there was a mood of optimism within the DZR as 1936 drew to a close. As Ernst Lehmann recorded in his memoirs, 'World traffic via airships has begun… it will spread out over all the seas and continents. The world should be grateful to Germany as the trailblazer.'

▲ *Published in March 1937, a DZR publicity brochure for the South America services.*

Zeppelin Cars

Between the two wars there was a fashionable fad among the motoring elite on both sides of the Atlantic to own opulent cars built with aircraft or even airship engines. Maybach, founded in 1909, was originally a subsidiary of the Zeppelin Company and became known for its superb diesel and gas engines developed for the airships and for rail cars. It began building a series of luxurious motor cars in 1919, and these included the 1930 DS7 Zeppelin, which was powered by a 6967cc V12 putting out 150hp at 2800rpm. These engines make a wonderful deep burbling rumble, but while the V12 might sound like an airship, this heavy vehicle drives more like a lorry and today's enthusiastic owners require an HGV licence before they can go for a spin. Top speed is 90–93mph (145–150km/h) and around 200 DS7 and DS8 were built.

Production of Maybach cars did not resume after the Second World War, although the marque was revived in the early 1990s. These highly prestigious cars are now marketed alongside Mercedes-Benz and the company is owned by Daimler AG.

Beautiful in black, a superb Maybach-engined Zeppelin car.

Of all airship crashes *Hindenburg*'s remains the most mysterious and most contentious. Many theorists were attracted to the idea of sabotage… But not only did the American investigators fail to find any evidence of sabotage, the Gestapo investigation was equally negative.

Len Deighton, *Airshipwreck*

On the evening of 6 May 1937, one year to the day since her first flight to the USA, the LZ129 *Hindenburg* had been due to arrive at the Naval Air Station at Lakehurst once again. This wasn't the first transatlantic crossing of 1937 as the airship had already made a roundtrip to South America in March, to be closely followed by the LZ127 *Graf Zeppelin*. By this time the smaller *Graf* was servicing the South Atlantic routes on a regular basis, while the *Hindenburg* was scheduled to make a total of eighteen flights to the USA and back.

On board the *Hindenburg* were thirty-six passengers, a number which fell far short of the maximum capacity which had been increased to eighty over the winter, plus a crew of sixty-one. Many of the crew members were undergoing training for the *Hindenburg*'s sistership, the LZ130, which was already under construction. As a result, there was a surfeit of officers crammed into the control car, although Hugo Eckener was not present and in command was Captain Max Pruss.

Did you know?
The Empire State Building has an airship mooring mast at its top, but no airship ever docked with it. Its primary purpose was to make the building that bit taller than the rival Chrysler Building.

▲ *The LZ129* Hindenburg *at the new international facility at Frankfurt. Two airship hangars were constructed at this location to house the* Hindenburg *and its sistership, the LZ130.*

Assisted by favourable winds, the *Hindenburg* appeared over the skyscrapers of New York three hours ahead of schedule. But upon reaching the airfield at Lakehurst, Pruss was advised that their early arrival meant that the landing crew and officials were not yet in place. Furthermore, a weather front approaching from the west was threatening to bring rain and thunderstorms to the area. Accordingly, he took the ship several miles to the south-east with the intention of sitting out the weather front and allowing time for the crew and officials to be mustered. The thunderstorms lingered in the vicinity for some time and it was around 7.00 p.m. when Pruss finally circled above the landing field to take a closer look at the surface conditions. Adjusting the airship's trim, he then made an approach, facing into the wind and descending to about 200ft (60m) and 700ft (210m) or so away from the mooring mast. The engines were throttled back leaving the silvery airship hanging almost motionless against the leaden sky.

As most of the passengers assembled at the large promenade windows to look for their families and friends waiting for them

crew. Moments later several observers on the ground spotted a glow and then a burst of flame just forward of the upper fin. Many said they heard a sound like the 'pop' of a gas stove. The flames spread fast from the rear of the ship, devouring the hydrogen cells and the hull at an incredible speed and sending up a gigantic mushroom cloud of fire and smoke.

The officers in the control car realised that something was wrong when a shudder ran through the airship and they saw the deadly orange glow reflected in the window panes. Losing buoyancy, the tail fell towards the ground, pointing the nose skywards at first. The fire tore through the passenger accommodation and a blast of flame shot up through the central axial corridor as if it was a chimney. Watching from the nearby hangars, radio reporter Herb Morrison reported the

◀ American Airlines poster for the Hindenburg's service to Europe. The airline provided air links to further destinations within the USA.

below, ropes were dropped from hatches near the airship's nose down to the ground

➤ *In a blaze of fire the* Hindenburg *falls to the ground at Lakehurst, New Jersey, on the evening of 6 May 1937.*

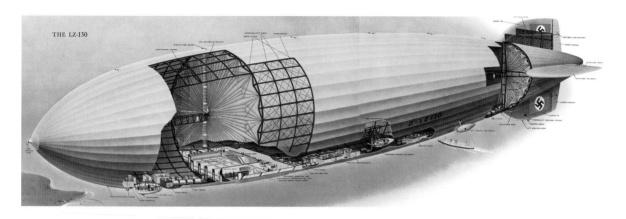

THE LZ-130

unfolding tragedy with undisguised horror. 'There's smoke and there's flames now... crashing into the ground – not quite to the mooring mast. This is terrible. This is one of the worst catastrophes in the world... Oh the humanity!'

Despite the inferno there were many miraculous escapes that day. Some passengers jumped from the airship's windows, while others fought their way out through the tangle of girders once the nose section had fallen to the ground. Sixty-two passengers and crew survived, while thirteen passengers and twenty-two members of the crew died in the fire or as a result of their injuries. This number is relatively small in comparison with the scale of modern airline disasters.

▲ *Despite the loss of the* Hindenburg, *the Zeppelin Company went ahead with plans for the LZ130 and this cutaway was published by* Fortune Magazine *in the USA.*

89

▲ *The LZ130* Graf Zeppelin *was almost identical to the* Hindenburg *apart from the arrangement of the passenger decks and forward facing engine pods to accommodate the water recovery gear.*

it was the burning of the hydrogen that had proved so devastating, but the official enquiry, or enquiries as the Germans held their own, failed to come up with a conclusive answer as to the source of its ignition. Sabotage has proved a popular and sensational theory but there is no hard evidence. It is more likely that a snapped bracing wire may have ignited escaped hydrogen collecting under the upper cover, or even sliced through one of the gas cells. More recently a rocket scientist in America put forward a theory that the special coating applied to protect the outer hull could have spontaneously ignited under certain atmospheric conditions.

This was, however, the first great disaster to be captured on film and, combined with the soundtrack of Herb Morrison's poignant commentary, the fall of the *Hindenburg* has been seared into the public consciousness for all time.

The cause of the accident has inspired much speculation over the years. Clearly

Construction work on the *Hindenburg*'s as yet un-named sistership, LZ130, was well advanced by the time of the disaster at Lakehurst, and there was every expectation that the Zeppelin Company would continue its transatlantic services. Hugo Eckener knew that this depended on the airship being inflated with non-flammable helium and the design of the LZ130 was modified accordingly. Helium is slightly heavier than hydrogen, so considerable weight savings had to be made. Helium is also very expensive and more difficult to replenish, therefore measures were needed to avoid any unnecessary venting and these included the installation of systems to recover water from the engine exhausts. But the main problem with helium was that the Americans held the major supplies of the gas which was extracted from natural gas deposits, mostly in Texas.

Completion of the new airship had been expected for late 1937. Publicity photographs were taken and brochures printed to attract bookings. Certainly the photographs reveal modernistic accommodation, although varying slightly in layout from the *Hindenburg*, with a dining room running across the width of the ship, and lounges and cabins arranged on either side.

▲ *Light and modern, the galley of the new LZ130 Graf Zeppelin.*

➤ *Inside one of the engine pods, an engineer monitors the performance of his Daimler-Benz DB602 diesel.*

The LZ130 would never carry a single fare-paying passenger because the Americans reneged on their promise to supply helium in the light of increased political tensions within Germany. The new airship finally flew for the first time on

14 April 1938, christened as the *Graf Zeppelin* (the original LZ127 being laid up in a hangar in Frankfurt). Inflated with hydrogen, the LZ130 made a number of propaganda flights and even probed the British radar defences strung along the eastern coast, but she never flew the Atlantic. Hermann Göring had both *Graf Zeppelins* dismantled and in the spring of 1940 the hangars at Frankfurt were dynamited to make way for the Luftwaffe's aircraft.

➤ *This nose cone on display at Friedrichshafen is almost all that remains of the LZ130 which was scrapped on the orders of Hermann Göring.*

The reason for the airship's great efficiency against submarines is obvious. Unlike the aeroplane, it can accommodate its speed to that of the slowest freighter, giving it constant protection if required. At this slow rate of travel airship crews have detected a submarine lurking as deep as 90ft below the surface of the water.

'Trail Blazing in the Skies', Goodyear Tire & Rubber Company, 1943

▼ Beta 1 was one of the first non-rigid or pressure airships evaluated by the British Army. She had begun life as Baby in 1909, but was enlarged to 33,307cu ft the following year.

BETA. THE ARMY AIRSHIP.
F. SCOVELL. PHOTO.   No 1.

After the fall of the giant rigids it was left to the lowly pressure airship or blimp to maintain a lighter-than-air presence. The most obvious characteristic of the pressure airship is that it has no rigid framework, and instead the shape of its envelope is achieved by maintaining the internal pressure at levels slightly above the ambient outside air pressure. Because air pressure fluctuates according to a number of factors, most notably reducing with height, and the internal gas pressure will also vary depending on the effects of heating and so on, a mechanism is required to keep the envelope from either sagging or bursting. The solution is the 'ballonet' – the word

when it decreases. Most designs, but by no means all, have one ballonet towards the front of the envelope and one aft, and by adjusting the balance between the two it is possible to adjust the trim of the airship. It is not the case that the ballonets are used to make the airship climb or to point the nose up or down, as that function is accomplished using the elevators on the tail.

As mentioned in the introduction, blimps were flying before Count Zeppelin's progeny became airborne and they also attracted the interest of the military for aerial reconnaissance duties. Given their apparent vulnerability the blimp might seem an unlikely recruit, and yet these gas-bags provided robust and reliable service throughout both world wars. In the First World War it was the British in particular who deployed blimps on convoy escort and

◄ A photographic montage showing the main classes of airship operated by the US Navy during the Second World War. The L-ships were based on Goodyear's pre-war advertising blimps.

comes from the French for a small balloon. All pressure airships contain one or two air-filled ballonets which can be depleted when the gas pressure increases, or filled with air

anti-submarine patrols, most notably over the Channel and the North Sea. In total, 213 were built for the Royal Naval Air Service (RNAS), the majority being the SS-class and the bigger Coastals, C Stars and North Sea classes which featured a distinctive tri-lobe envelope developed by the Astra-Torres Company. The specifications for these varied throughout the course of the war. For example, the early SS airships had a relatively small envelope volume of 70,000cu ft (19,810cu m) and a basic open gondola adapted from an aircraft fuselage, while the final North Sea craft were 36,000cu ft (10,190cu m) and had a larger enclosed gondola.

During the Second World War it was the Americans who played the lead role. Shocked into action by Japan's attack on Pearl Harbor on 7 December 1941, the US Congress approved the '10,000 Plane Program' which included the procurement of airships for the US Navy. The Goodyear Company conducted the lion's share of

◄ By the early 1960s the non-rigid had reached its ultimate expression in the form of the massive ZPG-3W which carried internal radar dishes for Airborne Early Warning (AEW) duties.

the manufacturing work, and a large-scale training programme also swung into action. Several of Goodyear's advertising airships were immediately pressed into service, mainly of the L-ship class, and they served as training ships or on maritime patrols until the bigger and purpose-built K-ships came on line.

The K-ships were without doubt the stalwarts of the US Navy's wartime airship force. Their specifications varied slightly and saw the envelope size reach 425,000cu ft (12,028cu m) by the end of the war. Powered by two air-cooled engines, mostly Pratt & Whitney 425hp radials mounted on outriggers, they had a range of over 2,000 miles (3,200km), an endurance of thirty-eight hours and a cruising speed of around 58mph (93km/h). The crew of ten consisted of pilots, navigator, ordnance-

man, mechanics and two radiomen. As well as escorting shipping, the K-ships had teeth with which to attack enemy submarines, including four Mk 47 depth charges and a Browning machine gun. Goodyear built 135 of the K-class airships and they were deployed over the Atlantic and Pacific Oceans and at several locations further afield. In 1944 the first of eight K-ships were flown across the Atlantic to operate from Port-Lyauty, in French Morocco, in order to carry out anti-submarine patrols in the Gibraltar Straits and, later on, to

◄◄ *In the decades following the Second World War, Goodyear's fleet of promotional blimps, including the* Mayflower *shown here, became familiar to thousands of American sports fans. (Goodyear)*

▼ *Elevation of a Goodyear blimp showing the internal arrangement of the ballonets, fore and aft, and the catenary curtain which supports the gondola from the top of the envelope.*

▲ *Goodyear's* Europa *N2A in her prime. Photographed at Bristol Airport in June 1984. Note the rows of small Nightsign bulbs along her flanks.*

assist in mine spotting and clearing in the Mediterranean.

During the war the US Navy's blimps provided a valuable and largely unrecognised service in protecting shipping convoys from the U-boat threat and rescuing downed airmen. But with the outbreak of peace most were moth-balled or scrapped,

although a few were sold off to become advertising blimps. An ever diminishing band of airship supporters within the Navy continued to fight their corner and development of the pressure airship peaked with the massive 1,516,300cu ft (42,911cu m) ZPG-3W Airborne Early Warning (AEW) airships. However, the arrival of all-seeing satellites meant that the writing was on the wall for the airships and the US Navy decommissioned its last ones in 1962.

Throughout the 1960s and 1970s the Goodyear Company famously operated its own small fleet of promotional blimps, making appearances above countless sports stadiums in the USA. In 1972 they were joined by the *Europa* which was assembled at Cardington, in the UK, and, based in Italy, she served as the company's European 'aerial ambassador'.

The 1980s may in fact prove to be the renaissance in airship operations and designs with many proposals are being expounded throughout the world.

Modern Transport magazine, 1978

While a handful of small pressure airships continued to fly, mostly in the USA, the dedicated supporters of the rigid airship were still dreaming of bigger things. During the Second World War the Goodyear Company continued to push for a new generation of rigids. When government backing was not forthcoming for a new fleet of flying aircraft carriers, Goodyear turned instead to industry and the general public for support through an advertising campaign depicting their vision for the passenger airship of tomorrow.

'Cruise the world in a flying hotel,' proclaimed the headline. 'We're talking about the greatest airship ever conceived. It will carry more than a hundred passengers in uncrowded comfort. It will take you for long cruises to many lands – on two-week vacation – and bring you home as relaxed and refreshed as you'd be from a stay in the finest resort hotel.' Goodyear's proposal was for a massive rigid airship, 50 per cent larger than the *Hindenburg*. With a helium volume of 10,000,000cu ft (283,000cu m), it would be 950ft (290m) long and have an operating range of over 6,000 miles (9,650km) – more than enough to fly to Europe and back again without stopping. As for the 'luxurious hotel-like comfort',

Did you know?

Metalclads were airships built with thin sheets of metal to form the gas-tight hull. Only four have been built and of those only two flew, including the US Navy's ZMC-2 which was known by her crew as the 'tin bubble'.

the colour illustrations depicted spacious state rooms and saloons featuring that old favourite, a grand piano. Inevitably they were fighting a losing battle against the upstart airliners that had benefited from the accelerated aeronautical progress of the war, in particular with regard to jet engines, and Goodyear's super-dirigibles remained no more than a dream.

In the 1950s the prospect of nuclear-powered aircraft gave the airship's proponents a fresh impetus. The US Air Force had commenced tests to develop nuclear power plants for bombers that could stay airborne for indefinite periods, cruising within range of intended targets. Although the programme was dropped in 1961, nuclear power was seen by many as a way of solving the airship's perennial problem of fuel displacement on longer flights. A number of highly fanciful schemes began to appear in the popular press, such as the 1,000ft (305m)-long 'Atoms

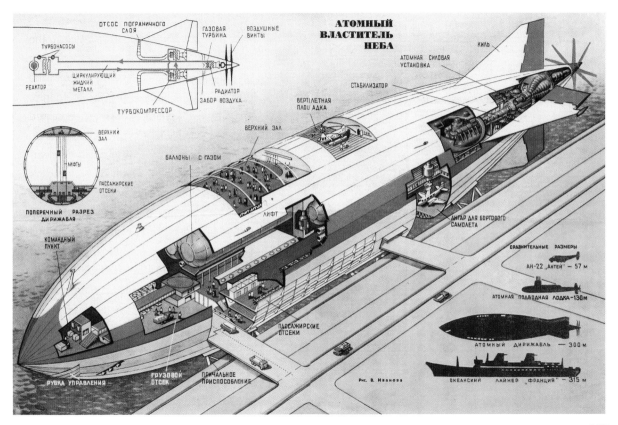

ОТСОС ПОГРАНИЧНОГО
СЛОЯ

ГАЗОВАЯ ТУРБИНА

ВОЗДУШНЫЕ
ВИНТЫ

**АТОМНЫЙ
ВЛАСТИТЕЛЬ
НЕБА**

КИЛЬ

ТУРБОНАСОСЫ

АТОМНАЯ СИЛОВАЯ
УСТАНОВКА

ЦИРКУЛИРУЮЩИЙ
ЖИДКИЙ
МЕТАЛЛ

РЕАКТОР

РАДИАТОР

СТАБИЛИЗАТОР

ЗАБОР ВОЗДУХА

ТУРБОКОМПРЕССОР

ВЕРТОЛЕТНАЯ
ПЛОЩАДКА

ВЕРХНИЙ
ЗАЛ

ВЕРХНИЙ ЗАЛ

ЛИФТЫ

БАЛЛОНЫ С ГАЗОМ

ПАССАЖИРСКИЕ
ОТСЕКИ

АНГАР ДЛЯ БОРТОВОГО
САМОЛЕТА

**ПОПЕРЕЧНЫЙ РАЗРЕЗ
ДИРИЖАБЛЯ**

ЛИФТ

КОМАНДНЫЙ
ПУНКТ

СРАВНИТЕЛЬНЫЕ РАЗМЕРЫ

АН-22 „АНТЕЙ" — 57 м

АТОМНАЯ ПОДВОДНАЯ ЛОДКА — 130 м

ПАССАЖИРСКИЕ
ОТСЕКИ

АТОМНЫЙ ДИРИЖАБЛЬ — 300 м

РУБКА УПРАВЛЕНИЯ

**ГРУЗОВОЙ
ОТСЕК**

**ПРИЧАЛЬНОЕ
ПРИСПОСОБЛЕНИЕ**

Рис. В. Иванова

ОКЕАНСКИЙ ЛАЙНЕР „ФРАНЦИЯ" — 315 м

103

➤ Goodyear's proposal for a heavy-lift airship which combined rotors with buoyant lift was published in the 1970s. (Goodyear)

➤ The 'Deltoid Pumpkin Seed' was the Aereon Corporation's proof-of-concept prototype for a lifting-body airship. Officially designated as the Aereon 26, it flew without helium in 1971. (Aereon Corporation)

for Peace Dirigible' published in *Mechanix Illustrated*. In the 1960s Professor Morse of Boston University gave the concept some credence with his designs for a large nuclear airship of 12,500,000cu ft (354,000cu m) which was designed to carry either cargo or up to 400 passengers. The nuclear power plant would drive three 4,000hp gas turbines turning two massive contra-rotating propellers situated in the tail, and a pair of 1,000hp turbofans.

Morse's utopian vision for a nuclear-powered airship was widely publicised, and even mimicked by the Russians who published artist impressions of their own version featuring helicopter decks, aircraft hangars and palatial accommodation on a scale that left precious little space for any helium. The Austrian engineer Erich von Veress went one better with a proposal for the AVL-1 with a volume of 14 million cu ft (396,000cu m) and accommodation for 500 passengers. That none of these designs ever saw the light of day has as much to do with the astronomical cost of building such a craft as with concerns about the safety of flying with nuclear reactors.

Meanwhile, other engineers had been focusing their efforts on augmenting the lifting power of the airship by combining the aerostatic lift of helium with aerodynamic lift generated by a wing or an aerofoil shaped hull, or alternatively through the addition of helicopter-type rotors. This has led to a whole family of variants known as 'hybrids'. The use of rotors to create a heavy-lift airship was proposed by Goodyear, and also by the aeronautical pioneer Frank Piasecki, who constructed a flying prototype of his Heli-Stat design using four military helicopters attached to a surplus airship envelope. During test flights at Lakehurst in July 1986, the Heli-Stat's frame was shaken apart resulting in the death of one pilot.

This accident highlighted the complexities of marrying two technologies in one craft, although more recently Boeing has taken an interest in the airship-rotor concept as a means of transporting large loads to and from otherwise inaccessible mineral mines in the northern Canada.

▲ *A scale model of the Skyship saucer design. Flown inside one of the Cardington hangars in 1975, it was typical of a number of proposals for lenticular airships.*

▲ *Proposed in the 1980s, the curious looking Magnus MA-32 featured a rotating envelope which generated additional lift through the Magnus effect. It never progressed beyond the scale model stage. (Magnus)*

There have also been numerous attempts to build a hybrid known as a lifting body – an airship which gains aerodynamic lift by virtue of its shape. In 1970 the US-based Aereon Corporation succeeded in test flying an experimental craft shaped like a pumpkin seed, the Aereon 26, although larger production versions did not follow. Another shape guaranteed to capture the imagination is the lenticular airship, or flying saucer. These have been proposed for some time and became particularly popular with airship designers in the 1970s. In the UK, Thermo-Skyships proposed a lenticular airship with a 10-ton lifting capacity, leading on, if successful, to a 25-ton version. A radio-controlled scale model was demonstrated at Cardington in 1975, much to the amusement of the gathered press who labelled it the 'Zeppsaucer'. The saucer concept has been taken up by designers in other countries, most notably in Mexico and Russia, but a full-size craft has yet to fly.

One of the advantages claimed for the lenticular airship is the ability to travel in any direction without the need for a tail and rudder, and this notion has led some to explore totally spherical designs. The Magnus airship of the 1980s resembled a bat-like creature, with the ball-shaped envelope on its back. This would be rotated horizontally to gain additional lift through the Magnus effect, created when a spinning object experiences a force perpendicular to the line of motion. A number of prototype spherical airships have also been developed by 21st Century Airships of Canada, which incorporate the control cabin within the base of the sphere itself.

◄ *A manned spherical airship built and flown by Hokan Colting of 21st Century Airships in Canada. The flight deck area is contained within the envelope.*
(21st Century Airships)

Can the airship make a comeback as a practical transport vehicle? Passenger travel by airship may sound somewhat too relaxed for the businessman in a hurry to get somewhere. However, the strongest argument for the comeback of the airship is not for passenger travel, but for cargo transport.

Future Life magazine, December 1979

In one of the most unusual share offerings ever seen, a German company that plans to make giant airships is launching on the Frankfurt Stock Exchange... The company, CargoLifter AG, intends to reinvent the Zeppelin as a modern form of freight transport. The airship will be so long, say its designers, that it will be able to lift the equivalent of ten fully-loaded trucks.

BBC News, May 2000

Did you know?

Proposals have been made to use robotic airships, or 'aerobots', for the exploration of those planets or moons that have an atmosphere thick enough to provide for buoyant flight

Despite a plethora of grand schemes featuring giant airships and an ever increasing variety of hybrids in all shapes and sizes, the design and construction of the pressure airship had little changed for decades by the 1970s. Indeed, the Goodyear airships were flying with reconditioned gondolas from the US Navy's wartime training ships. That was all about to change thanks to the efforts of a young naval architect working in Britain.

Roger Munk became fascinated with airships after reading about the R101 and he set about reinventing the pressure airship by

◄ Developed by Airship Industries in the 1980s, the Skyship 600 features fly-by-wire flight systems and vectored thrust for improved manoeuvrability.

giving it a modern twist. The Skyship series featured many elements that have become standard for the new generation of airships: new lightweight materials for the envelope, gondola and control surfaces, fly-by-wire controls which activate servo motors at the control surfaces, twin yoke control columns instead of the outmoded heavy foot pedals and elevator wheel working by direct mechanical means and, perhaps

▲ A Skyship 600 operating passenger sightseeing flights in Europe. (Skycruise Switzerland)

▲ ➤ A Lightship A60+ on the mast in front of the airship sheds at Cardington, prior to a European promotional tour.

most importantly, a system of tilting or vectoring the propellers through an arc of up to 200 degrees to direct their thrust. The latter gave the Skyships unsurpassed manoeuvrability, especially when it came to take-off and landing.

The prototype for the Skyship series, known as the AD500, was built by Aerospace Developments in the giant airship sheds at Cardington and first flew on 3 February 1979. Unfortunately it was wrecked at the mast in severe gales shortly afterwards, but Munk's team regrouped as Airship Industries to build the Skyship 500 and 600 series throughout the 1980s. The SK500 was 170.5ft (51.8m) long and had a volume of 182,000cu ft (5,150cu m). It was powered by onboard Porsche engines which gave a respectable speed of up to 70mph. The stretched version, the SK600, was 193.5ft (60m) long with a volume of 235,400cu ft (6,662cu m) and could accommodate up to twelve passengers.

With their all-white envelopes, modernistic gondolas and state-of-the-art flight systems, the Skyships certainly looked the business, and at their peak they were operating in locations throughout the world. Their main roles were to be tourism, advertising and airborne surveillance, but by the end of the 1980s Airship Industries was struggling to find buyers for the airships. Thirty years on, a number of SK600s continue to operate under the ownership of Airship Management Services, and several have been upgraded with new Textron Lycoming IO-540 engines.

One of the drawbacks with large pressure airships is the cost of purchase and operation. In the USA, Jim Thiele and

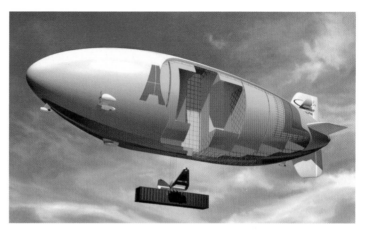

his American Blimp Company came up with the Lightship series as a cheaper alternative aimed specifically at the advertising market. The first Lightships put into production, the A60+ series, were less than half the volume of the Skyships at only 68,000cu ft (1,900cu m) and 128ft (39m) long. It was argued that size didn't matter as all airships look

▲ *Artist's impression of the proposed CargoLifter CL160 heavy-lift semi-rigid. The design changed in many details but it never progressed beyond the drawing board. (CargoLifter)*

big in the air. The Lightship also features a unique way of attracting attention as its translucent envelope enables it to be lit up internally to create the world's largest flying lightbulb which, incidentally, has resulted in a trail of UFO sightings whenever one flies in a new area.

In many ways the Lightships represent a retrospective step in terms of design. The gondola, suspended via external patches attached to the envelope, is very basic with traditional controls, plus there is no means of vectoring the engines. Having said that, the Lightships came to dominate airship advertising throughout the 1990s, and larger A150 and A170 versions have been produced for tourism and for evaluation by the US Navy.

It is interesting that the military, especially in the USA, continue to dabble with airships and in 2005 the Walrus project held the promise of a massive rapid response airship capable to transporting 500 tons direct to any location in the world, without the infrastructure required by conventional transport aircraft or the time limitations of sending men and equipment by sea. Several airship companies are continuing to explore

◀ *The prototype Zeppelin NT07 first took to the skies in September 1997. It features an internal framework with three longitudinal girders and three vectorable engines for unequalled manoeuvrability both in the air and on the ground.*

◀ *Close-up view of the NT07's passenger gondola. Note the absence of engines which are mounted higher up on the side of the envelope.*

➤ On board the Zeppelin NT07 003, the pilot points out notable landmarks during a hop across the Channel from the UK to Brussels in 2008.

➤ When not in the air, the Zeppelin NT07 can be docked to this hefty mobile mooring mast and is allowed to weathervane to keep it facing into the wind.

the lifting-body concept. Roger Munk's team (re-branded as Airship Technologies Group and more recently as Hybrid Air Vehicles) has produced designs for a series of heavy-lift airships known as SkyCats. Unfortunately, funding for large military projects is at the whim of the politicians and Walrus never made it off the drawing board, but HAV and a number of other companies, including Lockheed Martin

in the USA, are developing lifting-body designs. Lockheed Martin successfully flew their P-791 manned prototype in 2006.

In Europe, interest in the potential commercial market for a heavy-lift airship

◄ *Inside the hangar at Friedrichshafen, a Zeppelin NT07 undergoing a routine maintenance inspection. Note the inverted 'Y' tail configuration and rear-mounted engine.*

➤ *The Z-ship is a proposal for a larger forty-person Zeppelin, based on the NT design, proposed by the Zeppelin Europe Tourismus organisation. (ZET)*

© 2005 gk4. medienwerkstat

that can transport large or indivisible loads to any location saw the creation of the CargoLifter Company in the 1990s. Based at a former East German fighter airfield to the south of Berlin, CargoLifter proposed a gigantic semi-rigid airship twice the size of the *Hindenburg*, known as the CL160, capable of lifting loads of up to 160 tons. Their work attracted considerable interest at the time and a vast jelly-mould hangar, to house two CL160s, was constructed at the Briesen-Brand site. A small one-man test airship and even a heavy-lift spherical balloon did fly, but the CL160 failed to

materialise and CargoLifter went belly-up in 2002. The hangar is now an indoor tropical holiday resort.

Meanwhile a slumbering giant was stirring. By the 1990s the Zeppelin Company was ready to return to the lighter-than-air business with its own take on the modern airship. The Zeppelin NT07 features a rigid framework with three longitudinal girders joined by triangular transverse frames which provide the supporting points for the engines and the control surfaces.

◄ *There have been several designs for a high-altitude unmanned airship for telecommunications and AEW surveillance purposes. The StratSat was a short-lived British concept, while in the USA Lockheed-Martin is nearing completion of its High Altitude Long Endurance Demonstrator (HALE-D).*

It would be most accurate to describe this configuration as a semi-rigid because the outer envelope, although attached to the framework, still requires internal gas pressure to maintain its shape. The NT07's 290,000cu ft (8,255cu m) volume and length of 246ft (75m) make it the largest manned airship operating at present. The framework provides attachment points for the Textron Lycoming IO-360 engines, vectorable of course, positioned well away from the gondola to reduce cabin noise, plus a third one mounted at the tail. The result is an airship with vastly improved ground manoeuvring capabilities which drastically reduces the need for large ground crews.

The prototype NT07 first flew in 1997. Three more have been built since then and are operating in Germany, Japan and the USA. More are sure to follow and a group called Zeppelin European Tourismus (ZET) hopes to establish an airship shuttle service between major European cities using a larger version of the Zeppelin.

Although the airship might not be the solution to all our transport needs, its unique flight characteristics make it ideally suited to a number of roles. To predict a renaissance of the airship is to overstate the case, but airships are being taken seriously again. There is interest in heavy-lift airships to serve inaccessible regions of northern Canada, and the US military is proceeding with development of a colossal unmanned high-altitude airship for communications and surveillance missions. Keep watching the skies.

One only has to have the will for it to succeed.

Ferdinand Graf von Zeppelin, 1838–1917

ZEPPELINS (GERMANY)

Note that although each airship built by Zeppelin was given a works number, the German Imperial Army and Navy both applied their own numbering systems for the airships they operated.

Design type	Zeppelin work's number	First flight	Length ft (m)	Volume cubic ft (cubic m)	Max speed mph (km/h)
'a'	LZ-1	1900	420 (128)	399,000 (11,300)	17 (27)
'b'	LZ-2 LZ-3/LZ-3A	1905 1906	420 (128) 420-446 (128-136)	399,200 (11,300) 399,200-430,800 (11,300-12,200)	24 (38) 33 (53)
'c'	LZ-4, LZ-5	1908-1909	446 (136)	529,725 (15,000)	30 (48)
'd'	LZ-6/LZ-6A	1909	446-472.5 (136-144)	529,725-565,000 (15,000-16,000)	34 (54)
'e'	LZ-7 *Deutschland*, LZ-8 *Ersatz Deutschland*	1910-1911	485.5 (148)	681,580 (19,300)	37 (59)
'f'	LZ-9, LZ-10 *Schwaben*, LZ-12	1911-1912	459 (140)	628,600 (17,800)	47 (76)
'g'	LZ-11 *Viktoria Luise*, LZ-13 *Hansa*	1912-1913	485.5 (148)	660,000 (18,700)	51 (82)
'h'	LZ-14 15-16, LZ-17 *Sachsen*, 19-20	1912 1913	518 (158) 466 - 459 (142-140)	797,120 (19,740) 688,625 (19,500)	47.5 (76)
'i'	LZ-18	1913	518 (158)	953,350 (27,000)	47 (76.5)
'k'	LZ-21	1913	485.5 (148)	737,000 (20,870)	46 (74)
'l'	LZ-22, LZ-23	1914	511 (156)	781,800 (22,140)	45 (72)

Design type	Zeppelin work's number	First flight	Length ft (m)	Volume cubic ft (cubic m)	Max speed mph (km/h)
'm' L3 class	LZ-24, LZ-25, 27-35, LZ-37	1915	518 (158)	793,530 (22,470)	52 (84)
'n'	LZ-26	1914	528 (161)	880,000 (25,000)	50 (80.5)
'o'	LZ-36, LZ-37	1915	530 (161.5)	879,350 (24,900)	53 (85)
'p' L10 class	LZ-38, 40-58, LZ-60, LZ-63	1915-1916	536.5 (163.5)	1,125,485 (31,900)	60 (96.5)
'q' L20 class	LZ-59, LZ-61, 64-71, LZ-73, LZ-77, LZ-81	1915-1916	585.5 (178.5)	1,264,300 (35,800)	59 (95)
'r' L30 class	LZ-62, LZ-72, 74-76, 78-80, 82-90	1916-1917	649.5 (198)	1,949,390-1,942,325 (55,200-55,000)	64 (103)
's'	LZ-91, LZ-92	1917	644.5 (196.5)	1,942,325 (55,500)	64 (103)
't'	LZ-93, LZ-94	1917	644.5 (196.5)	1,970,580 (55,800)	65 (104.5)
'u'	LZ-95, 96-99	1917	644.5 (196.5)	1,970,580 (55,800)	70 (112.5)
'v' L53 class	LZ-100, LZ-101, LZ-103, 105-111	1917-1918	644.5 (196.5)	1,977,640 (56,000)	70 (112.5)
'w'	LZ-102, LZ-104	1917	743 (226.5)	2,419,000 (68,500)	64 (103)
'x' L70 class	LZ-112, LZ-113, LZ-114 (Dixmude)	1918-1920	692-743 (211-226.5)	2,196,600-2,419,080 (62,200-68,500)	81 (130)
'y'	LZ-120/LZ-120A Bodensee (Esperia) LZ-121 Nordstern (Mediterranée)	1919 1921	396-429 (120.8-130.8)	776,930-796,350 (20,000-22,550)	82 (132)
'z'	LZ-126 (ZR3 Los Angeles)	1924	656 (200)	2,472,050 (70,000)	73 (117.5)

Design type	Zeppelin work's number	First flight	Length ft (m)	Volume cubic ft (cubic m)	Max speed mph (km/h)
-	LZ-127 *Graf Zeppelin*	1928	776 (236.6)	3,700,000 (105,000) hydrogen/Blaugas	79.5 (128)
-	LZ-129 *Hindenburg* LZ-130 *Graf Zeppelin* (II)	1936 1938	804 (245) 84 (245)	7,063,000 (200,000)	

NT07*	ZLT NT07 001-004 (to date)	1997-2008	245 (75.5)	291,500 (8,255)	78 (125)

*Strictly speaking the NT07 should be classed a semi-rigid as it depends upon a combination of girders and internal pressure to maintain its shape.

OTHER GERMAN RIGID AIRSHIPS

Type	Airship	First flight	Length ft (m)	Volume cubic ft (cubic m)	Max speed mph (km/h)
-	Schwarz No.2 (metalclad)	1897	159 (47.5)	13,066 (3,700)	17 (27)
'a'	Shütte-Lanz SL.1 (wooden frame)	1911	430 (131)	723,958 (20,500)	44 (71)
'b'	Shütte-Lanz SL.2 (wooden frame)	1914	472.5-512 (144-156)	882,875-971,165 (25,000-27,500)	55 (88.5)
'c' SL.3 class	Shütte-Lanz SL.3, 4-5 (wooden frame)	1915	502 (153)	1,144,560 (32,410)	52.5 (84.5)
'd' SL.6 class	Shütte-Lanz SL.6, SL.7 (wooden frame)	1915	531.5 (162)	1,240,616 (35,130)	58 (93)

Type	Airship	First flight	Length ft (m)	Volume cubic ft (cubic m)	Max speed mph (km/h)
'e' SL.8 class	Shütte-Lanz SL.8, 9-19 (wooden frame)	1916-1917	570 (174)	1,240,616-1,369,516 (35,130-38,780)	60 (96.5)
'f' SL.20 class	Shütte-Lanz SL.20, 21-22 (wooden frame)	1917-1920	649.5 (198)	1,977,640 (56,000)	65 (104.5)

BRITISH RIGID AIRSHIPS

Type	Airship	First flight	Length ft (m)	Volume cubic ft (cubic m)	Max speed mph (km/h)
-	HMA No.1 'Mayfly'	1911	512 (156)	663,500 (18,788)	42 (19)
-	Vickers No.9	1916	530 (161.5)	889,300 (25,181)	45 (72.5)
23 class	Vickers No.23-25, R26	1917-1918	535 (163)	997,640 (28,250)	52 (84)
23X class	Beardmore R27, R29	1918	539 (164)	990,577 (28,050)	55 (88.5)
31 class	Short R31. R32	1918, 1919	614.5 (187)	1,555,000 (43,975)	71 (114)
33 class	Armstrong Whitworth R33, Beardmore R34	1919	643 (196)	1,958,553 (55,460)	60 (96.5)
36 class	Beardmore R36	1921	672.5 (205)	2,120,000 (60,032)	65 (105)
-	Royal Airship Works R38 (US Navy's ZR-2)	1921	695 (212)	2,740,420 (77,600)	66 (106)
-	Vickers R80	1920	531.5 (162)	1,260,000 (35,680)	60 (96.5)

Type	Airship	First flight	Length ft (m)	Volume cubic ft (cubic m)	Max speed mph (km/h)
-	Airship Guarantee Company R100	1929	709 (216)	5,156,000 (146,060)	81 (130)
-	Royal Airship Works R101	1929	720-731 (219 - 223)	4,998,500-5,509,090 (141,542-156,000)	70 (112.5)

AMERICAN RIGID AIRSHIPS

Type	Airship	First flight	Length ft (m)	Volume cubic ft (cubic m)	Max speed mph (km/h)
-	Naval Aircraft Factory ZR-1 *Shenandoah*	1923	680 (207)	2,151,200 (60,915)	62.5 (101)
-	Airship Development Corp ZMC-2 (metalclad)	1929	149.5 (45.5)	202,000 (5,720)	62 (100)
-	Slate Aircraft *City of Glendale* (metalclad – only flown tethered and unpowered)	1929	212 (64.6)	330,000 (9,345)	-
-	Goodyear Zeppelin ZRS-4 *Akron* Goodyear Zeppelin ZRS-5 *Macon*	1931 1933	6,850,000 (193,970)	785 (239)	84 (135)

Please note that specifications for individual airships vary with different sources of information and we have deferred to Putnam's *Zeppelin Rigid Airships*, published in 1992.

Although none of the great rigid airships have survived intact, there are still a number of locations providing the opportunity to connect with these bygone leviathans.

THE ZEPPELIN MUSEUM, FRIEDRICHSHAFEN, GERMANY

Friedrichshafen is Zeppelin town, and one of the main attractions is the award-winning Zeppelin Museum, located in the Bauhaus-style harbour building. The items on display include an original engine pod from the *Graf Zeppelin*, but undoubtedly the focal point of the museum has to be the full-size replica of a section of the *Hindenburg*. (www.zeppelin-museum.de)

Friedrichshafen is also home to the Zeppelin Company and although the old hangars are long gone a new one has been constructed beside the airfield for the construction of the new Zeppelin NT airships. Flights and hangar tours are available. You can also fly in San Francisco or Tokyo, and occasionally at other locations when the Zeppelins are on tour. (www.zeppelintours.com)

ZEPPELIN MUSEUM, MEERSBURG, GERMANY

Just down the road from Friedrichshafen is the independent Zeppelin Museum at Meersburg which features the collection of Heinz Urban. (www.zeppelinmuseum.com)

ZEPPELIN AND GARRISON MUSEUM, TONDOR, DENMARK

During the First World War the German Imperial Navy operated Zeppelins from an airbase at Tondor, now part of Denmark.

Although the hangars are long gone, this small museum is of interest.
(www.zeppelin-museum.dk)

RAF MUSEUM, LONDON, UK

Among the aeroplanes displayed at Hendon is a section of the R33's forward control car.
(www.rafmuseum.org.uk/london)

CARDINGTON, BEDFORD, UK

This is a very special place for the airship-minded. The two massive hangars, or sheds, stand as poignant relics of the Imperial Airship Scheme. This is where the Skyship series was developed and built. There is some redevelopment around the site and the former office building for the Royal Airship Works still stands; across the A600 is Shortstown, which once housed the workers. Nearby, in the village cemetery at Cardington, there is the mass grave of the forty-four men who perished in the R101, and the airship's flag hangs in the church. At the time of printing there is no public access to the Cardington hangars.

US AIRSHIP HANGARS

Although many of the original hangars have been lost, there are several survivors, including those at Lakehurst in New Jersey, Moffett Field (formerly Sunnyvale) in California and the former Goodyear construction hangar at Akron, Ohio. In addition there are a number of Second World War hangers. It should be noted that many of these facilities are either on military bases or in commercial use and may not be accessible to the public. One exception is at the former Naval Air Station at Tillamook, Oregon, now home to the Tillamook Air Museum.
(www.tillamookair.com)

Aerodynamic lift   Lift obtained by the curved shape of an airship pushing through the air, in the same way that an aircraft's wing creates lift.

Ballonets   Expandable air-filled chambers within the envelope of a pressure airship, used to regulate internal pressure.

Blimp   Colloquial term favoured by the Americans for a pressure airship.

Duralumin   A family of alloys of aluminium.

Envelope   The main part of a pressure airship which contains the gas and ballonets.

Gas capacity   The volume of gas contained at 100 per cent full.

Gas cells   These contain the lifting gas within a rigid airship.

Gondola   A boat-shaped compartment, hence the term, suspended beneath an airship. Often referred to as a car by in the US.

Gross lift   The total lift of the gas contained in an airship.

➤ *This impressive full-scale reconstruction of part of the LZ129* Hindenburg *is at the Zeppelin Museum in Friedrichshafen, the historical home of the Zeppelin Company.*

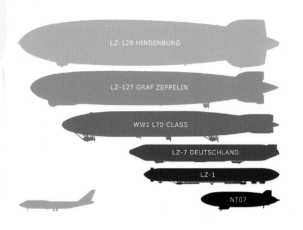

◄◄ *The comparative scale of various Zeppelins, including the latest NT07, is indicated in this chart alongside a Boeing 747 Jumbo Jet.*

◄ *The USS* Macon *in the hangar at Sunnyvale, California. This former US Navy facility is now the location for the Airship Ventures passenger operations with the Zeppelin NT07 004.*

Hangar	Buildings for airships, referred to as 'sheds' in Britain or '*halle*', meaning halls, by the Germans, i.e. *Luftschiffhalle* or *Zeppelinhalle*.
Hybrid	An airship that incorporates buoyant lift with other forms of lift, such as aerodynamic lift or through rotors.
Lifting gas	Any gas lighter than air which can be used to fill an airship. The most commonly used are hydrogen, helium and hot air.
Longitudinals	The main girders running lengthwise on a rigid airship.
Luftschiffe	German term for airships.
Main rings	The transverse rings on a rigid airship.

Non-rigid	Sometimes known as a 'pressure airship' or blimp – it maintains its shape through internal gas pressure.
Payload	The portion of the useful lift available to carry passengers or cargo.
Pressure airship	See non-rigid.
Rigid airship	Airship with a rigid frame.
Semi-rigid	A pressure airship with a rigid keel.
Thermal airship	Sometimes known as a hot-air airship, gains lift as with a hot-air balloon.
Useful lift	Amount of lift available after subtracting the weight of the airship from its gross lift.
Vectored thrust	The ability to direct thrust from an engine.

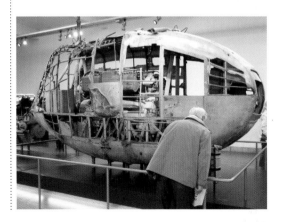

➤ *Engine pod from the LZ127 Graf Zeppelin, on display at the Zeppelin Museum in Friedrichshafen.*